DNA TO AFRICA

THE SEARCH CONTINUES

By

Aaron L. Day

Award-Winning & Best-Selling Author

Of

**Locating Free African American Ancestors:
A Beginner's Guide**

******

History Lessons

&

**The Heritage of African Americans In
Long Beach: Over 100 Years**

**Dedicated to the memory of
Professor John Hope Franklin 1915-2009**

DNA TO AFRICA

THE SEARCH CONTINUES

By

Aaron L. Day

ISBN 978-0-7414-5944-2

Printed in the United States of America

Published   March 2013

INFINITY PUBLISHING
1094 New DeHaven Street, Suite 100
West Conshohocken, PA 19428-2713
Toll-free (877) BUY BOOK
Local Phone (610) 941-9999
Fax (610) 941-9959
Info@buybooksontheweb.com
www.buybooksontheweb.com

DEDICATION

To the memory of Professor John Hope Franklin, 1915-2009

and

To the Day and Banks Family Members

Paternal and Maternal Families of Aaron L. Day

ACKNOWLEDGMENTS

In doing the research for and writing this book, I have many people to thank for their assistance. I would like to give a very special thank you to everyone who helped make this publication possible. Thanks to the work of Professor John Hope Franklin, Paul Heinegg, and others who have researched, documented, and preserved the history and heritage of African Americans. This book is dedicated to the memory of John Hope Franklin. His groundbreaking work in researching, documenting, and writing about the history of free African Americans, is greatly appreciated, and has helped me tremendously.   Thanks also to people like Gina Paige, and Dr. Rick Kittles, of African Ancestry who are pioneers in the DNA/Genealogy research field. Thanks to them, I was able to make the DNA to Africa connection with my ancestors.

A special thanks to my good friend Jane Buel Bradley, who passed away in 2002. I enjoyed her support and inspiration, and cherished her friendship. She gave me permission to include her poem 'Tree of Life' in 'Locating Free African American Ancestors: A Beginner's Guide.' She also gave me permission to use 'Unsmogged Martin Luther King Day,' which I have included in this book. Thanks to Natalie Pryor, for the information that she sent me regarding one of my ancestors, (Augustin Day). Her information was so helpful, and cleared up a big mystery. Thanks again to Nancy Carlberg, who published 'Locating Free African American Ancestors: A Beginner's Guide.' Her advice and help to me through the years is greatly appreciated.

This is the third book that Office Max IMPRESS of Lakewood, California has assisted me with the picture layouts. Thanks to Jon Rocha for all of his help. A very sincere thanks is given to everyone who submitted photos and materials for this project. I would especially like to thank the Mombasa/Long Beach Sister Cities delegation for their materials, which are included in the 'Discovery of My Homeland' chapter of this book. Bill Preston, Evelyn Knight, Jesse B. Johnson, Jr., Patricia Maye, Beverly Ragland, and Julie Hatter, made up the delegation from Long Beach to Mombasa, Kenya in November 2009. The information they shared with me about our Sister City, Mombasa is appreciated.

Also included in the 'Discovery of My Homeland' chapter are the interesting stories from my numerous friends from various parts of Africa, who are now living in the United States. Thanks to Issakson (Ghana), Cynthia (Cameroon), Edward (Ghana), Peter (Uganda), Mbaiceesay (Nigeria), and Peter (Camaroon), for sharing their stories. They all stressed the need to learn about the numerous cultures in the world, and the need for education.

I would like to give a big thanks to the Day and Banks family members who contributed information, and material for this book. The information received from my cousin Karin Berry, was so very helpful, and I am happy that she contacted me. I now have a connection to another side of the Day family. Again, thanks so much to my cousin Dora, for her help and inspiration, and for sharing her notes about our family history.

CONTENTS

PART I—FAMILY HISTORY RESEARCH

PART II—DNA TO AFRICA And THE DISCOVERY OF MY HOMELAND

ILLUSTRATIONS/PICTURES

* Cover designed by Infinity Publishing
* John Hope Franklin
* 'History Lessons' Book, by Infinity Publishing.com
* Aaron L. Day & (4 Generation Picture) – (Inset-Tyray)
* Emma Banks-Thompson & (4 Generation Picture)
* Various 'Banks' Family Reunions
* Olla Banks, grandmother of Aaron L. Day – Springfield Church Conference
* George H. Banks (poppa Banks), grandfather of Aaron L. Day, Aunt Emma & Uncle
    Chris Thomas – (Aaron – Second Grade Class Picture)
* Day Family, Sharon Day, Day Children
* William A. Day
* James G. Day, Jr.
* Aaron L. Day
* Theodore A. Day
* Frederick H. Day
* Sharon L. Day
* Day brothers & sisters, Day nephews
* Wedding Picture of Aaron L. Day
* Dad & family reunions –NC
* Day brothers & cousin Christine
* James Ware
* Helen Greaves
* William & Uncle Wilborn
* Mom, Aaron, & Poppa Banks
* Day brothers & Thompson sisters
* George & Eliza Banks
* Day family children
* Uncle Gus & first wife/Uncle Gus & Aunt Minnie
* William Howard Day
* 1880 U. S. Census Schedule
* 1870 U. S. Census Schedule
* 1860 U. S. Census Schedule
* 1850 U. S. Census Schedule
* 1840 U. S. Census Schedule
*. Boys Sawing Tree - Abby Aldrich Rockefeller Folk Art Center
* 1794 Tax List
* 1830 U. S. Census Schedule

* 1820 U. S. Census Schedule
* John J. Day, Jr., Chief Justice of Liberia/Map showing Liberia
* Thomas J. Day, Cabinetmaker -National Museum N. C.
* Thomas J. Day, Cabinetmaker - Union (Yellow) Tavern
* Dad's house – 3 pictures
* Day Family/Day women
* Locating Free African American Ancestors: A Beginner's Guide, Carlberg Press
* William Day receiving Military Awards
* Molecular Genealogy flyer
* African Ancestry (Liberia & Nigeria)
* Hausa Certificate
* KRU Certificate
* Charles R. Banks, Jr., DNA results
* Nigeria Map
* Peter Lobe
* Kenya Ports Authority/Long Beach Sister City delegation
* Camels and Bus delivering books
* Books presented to Mombasa by Long Beach Inaugural Delegation/Restaurant
* Likoni Aids Orphanage, Mombasa Kenya, & Mombasa Delegation in U.S.
* Evelyn's 75th Birthday & Tribal Group photo
* Mombasa Sister City delegation at Long Beach, California, Public Library
* Mombasa Sister City delegation with Mayor Bob Foster-City of Long Beach
* Mombasa Sister City delegation at Long Beach City Hall
* Four Grandchildren of Aaron L. Day, at Wilberforce University, in Ohio
* Naomi Rainey, NAACP Long Beach Chapter, President
* NAACP Book - Greene Graphics Studio
* AAHS-LB 13 Original Board Members
* Pedigree Ancestor Chart
* The Heritage of African Americans in Long Beach, by Infinity Publishing.com
* Discover Your Roots IV Conference

Also by Aaron L. Day

BOOKS

- **'The Eastonian' - High School Yearbook** – Xenia East High School, in Ohio, co-editor Mildred Scott, (1957).
- **Locating Free African American Ancestors: A Beginner's Guide** - Carlberg Press, Anaheim, California, (2003).
- **History Lessons** - Infinity Publishing, (2005).
- **Family Tree** - Greene Graphics Studio,   co-editor Naomi Rainey, (2006).
- **(Best-Seller) - The Heritage of African Americans in Long Beach: Over 100 Years** - co-editor Indira Hale Tucker, Infinity Publishing.com , (2007).
- **Preserving Our History: African Americans In Xenia, Ohio** - co-editor Theodore Day, InfinityPublishing.com - (June 2010).

AWARD-WINNING ARTICLES

* **Tracing My African American Ancestors** – (short story), Southern California Genealogical Society-National Writing Contest, (2001).

Also published in 'Celebrating Family History' an Anthology of Prize-Winning Stories published by the Southern California Genealogical Society, (2005).

 * **The Search for Free African American Ancestors** – (essay), Southern California Genealogical Society-National Writing Contest, (2002) also published in the Heritage Quest Genealogical Magazine, (2003).

 * **History Lessons** – (essay),   Iliad Press/Cader Publishing contest, (2002).

 * **What's Available For African American Research** – (advice & how-to article), Southern California Genealogical Society-National Writing Contest, (2004).

ADDITIONAL ARTICLES

* **The Values and Problems of Writing** – Report Writing Class, (1969).
* **The Unbelievable Supervisor** – Elements of Supervision Class, (1972).
* **Historical Comparison of: Revolt of 1831 and Riot of 1965** – (History Class, 1974).
* **Fun Packed Three Hours** - The American Family Good News Journal, (1993).
* **A Company Moral Dilemma** - History Lessons (book), (1994)
* **The Veteran in My Life** - Press Telegram Newspaper, (1998).
* **Mary Dell Butler** - History Lessons (book), (1999).
* **Thomas Day: Cabinetmaker** - Questing Heirs Genealogy Newsletter, (2000).
* **The System of Bonding** - Questing Heirs Genealogy Newsletter (2000).
* **The Lucky Miniature Tea Set** - Everton's Genealogical Helper Magazine, (2001).
* **They Are Still Free** - Friends of the Long Beach Public Library Newsletter, (2001).
* **About My Father** - Questing Heirs Genealogy Newsletter, (2002).
* **A Salute To Frances Curtis Bonds** - NAACP Long Beach Newsletter, (2002).

* **My Push For Civil Rights** - Library of Congress (Voices of Civil Rights Collection) (2004).
* **Civil Rights through the Years** - History Lessons (book), (2005).
* **The Story of the Day Brothers** - History Lessons (book), (2005).
* **The Journey: Early Childhood Memories** - History Lessons (book), (2005).
* **Rachel Day and the Day Family History** - History Lessons (book), (2005).
* **Ernest McBride, Sr. : Hero for Our People** - History Lessons (book), (2005).
* **Researching Free African American Ancestors – Pre - 1865** – Journal of the Southern California Genealogical Society, (2006).
* **DNA to Africa** - Everton's Genealogical Helper Magazine, (2006).
* **'Mary Dell Butler: Long Beach Community Leader'** - African American National Biography, Edited by, Henry Louis Gates, Jr., & Evelyn Brooks Higginbotham, (2008).

Contributed the Foreword to 'African-Americans In Hawaii: A Search For Identity,' by Dr. Ayin Adams, PhD. This book is edited by Dr. Adams, and is about the History and Heritage of African Americans in Hawai'i. President Barack Obama was born in Hawai'i, and his life leading up to, and including the election is featured in this book, (2010).

POEMS

* **Listening** - (1975).
* **By Invitation Only** - (1980).
* **I Want To Read** - (2003).
* **Carla** - (2003).
* **Dear Aunt** - (2003).
* **A Friend Named Anebel** - (2004).
* **I Know Who I Am** - (2005).
* **The Loaded Bookcase** - (2005).
* **For My Descendants** - (2005).
* **Advice to My Grandchildren** - (2005).
* **Ray M.** - (2005).
* **What Makes A Legend** - (2007).
* **What's Wrong With Trying** - (2009).

NOTE CARDS

* **History Cards** - (co-editor Kate Pedigo), set of 15 cards. Drawings are by Pedigo, stories are by Day, (2008).

Kate Pedigo, 98 year old artist and author, is affectionately known as 'Southern California's Grandma Moses.' Pedigo has published a total of four books.

Aaron L. Day is affectionately known as 'Mr. Library' because of his many years of volunteer work with the Long Beach Public Library. Day has published a total of four books.

This is the fifth project that Pedigo and Day have collaborated on. The two have hosted and produced several television shows together, and held numerous book signing events. Now they have created three sets of note cards. There are five note cards in the 'History Series,' five note cards in the 'Family Series,' and five note cards in the 'Inspiration Series.'

UNSMOGGED   Martin Luther King Day

Rising from dreams, I love to start slowly
On the day. But this morning
the wind comes smashing through the garden,
overturning pots
breaking off buds,
bruising flowers;
then sirens rush screaming
down the street.

I leap into the car,
drive fast to Signal Hill.
At the crest of the road
I see spread out before me
the full panoply of mountains,
usually lost in smog.

I get out, lean against the wind,
fill my body with air, rest my eyes
on the San Gabriel Range:
Mount Wilson, Old Baldy, San Gorgonio...
Horizon curves up and down,
slants to blue shadows
where mists blur San Jacinto.

Behind me I hear the creak and whir
of rocking horse drills, pulling up oil
from this historic spot, and the smack
of car doors as others come to see
the view on this clear day. One cries
"I can see where I work in L.A.!"
and I laugh to find all those
self-important skyscrapers
are mere pegs in a cribbage board
at the foot of the great mountains.

Today, we celebrate the birth
of a man who knew about mountains.
And the wind
has given them back to us.

By Jane Buel Bradley

History Lessons Book
By
Aaron L. Day

Infinity Publishing, 2005

PART I

FAMILY HISTORY RESEARCH

East High School - attended by the Day family in Ohio

Aaron L. Day & Mildred Scott – Co-Editors of

'The Eastonian' – High School Yearbook

INTRODUCTION

Writing this book has been an exciting, and very rewarding experience. It has also been very painful and emotional at times. Although, my 'Day ancestors were free in the deep-South during the institution of slavery, there were many hardships that they had to endure. Some of the language used to describe both the slaves, and the free people of color during the days of slavery was very harsh, and demeaning.

It is difficult to imagine someone being considered free, and having to spend 31 years of his/her life as an indentured servant, and sometimes, many years longer. Like the slaves of the time, they were also limited in their freedom. Laws were constantly being made that would limit the freedom and activities of the free people of color, as well as the slaves.

When I began this project, I had no idea that I would be traveling so far back in time (over three-hundred years). So much of what I have learned about the history of the free people of color was not available to me in school. To this day, many people are not aware of the volumes of records that are available, that detail the lives of our ancestors.

I have been asked a few times, "why spend so much time digging into the past?" I believe that if we are aware of some of the problems, and mistakes of the past, we don't run the risk of repeating them. I also believe that it is good to know ones heritage.

The research and writings of **Historian John Hope Franklin** have been very inspirational and helpful to me. Although I was born in Ohio, I have done a tremendous amount of research in North Carolina, the home of my father. Discovering Professor Franklin's book **'The Free Negro In North Carolina 1790-1860,'** at the beginning of my research was the perfect guide for me. In his foreword, he reviewed quite a few historians who had been involved in the study of blacks in America. Especially helpful to me, and to others who are interested in this part of our history, he listed the titles of their books. Franklin also talked about the research and writings of another pioneer in this field, **Herbert Aptheker**.

Franklin started out with **Carter G. Woodson,** and his two groundbreaking books: **'Free Negro Owners of Slaves in the United States in 1830,'** and **'Free Negro Heads of Families in the United States in 1830.'** I discovered many of my Day family members listed in both of these books. Franklin also listed three other historic books on the study of free blacks. In 1958, **Leon Litwack**, wrote **'North of Slavery: The Negro in the Free States, 1790 - 1860.** In 1974, **Ira Berlin**, wrote **'Slaves Without Masters: The Free Negro In the Antebellum South.** In 1981, **Leonard P. Curry**, published his **'The Free Black in Urban America, 1800-1850.'** For those interested in the study of free blacks, these books are excellent resources. Mr. Franklin included all of this helpful information at the beginning of his book. **'From Slavery To Freedom,'** also by Franklin, has been very useful, and is taught in many schools, throughout the world.

My ancestor research has also included the latest technology, DNA and Genealogy. When I began my family history research, this technology was not readily available. I am so happy I learned about the DNA connection, and that I began going to the different conferences that were available, so that I could learn more. The African Ancestry Company is a pioneer, and leader in this field. This is where I had my DNA tests done.

As explained in **'Locating Free African American Ancestors: A Beginner's Guide'** the primary purpose of that report was to review resources that may be used when researching ancestors who lived prior to the Civil War era. There is a wide range of resources available on free African Americans that may prove to be very useful to researchers. Many people are surprised to learn that about a half million, almost one-seventh of the Blacks of this country were free long before the end of the Civil War in 1865.

'Locating Free African American Ancestors: A Beginner's Guide' covers the following:

- The award-winning 'The Search For Free African American Ancestors' (the Day & Banks families).
- Beginning Your Research.
- What Resources Are Available?
- How To Locate Information About Free African American Ancestors
- What resources are available; genealogy books, magazines, websites, libraries, National Archives, Family History Centers, etc. It ends with the discovery of my 4-great-grandmother Rachel Day, in Caswell County, North Carolina.

In this report, **'DNA TO AFRICA: The Search Continues,'** The 'Day' family research continues from North Carolina into the State of Virginia, with the discovery of four additional generations - to the 1660's. This report reviews the discovery of my 7-great-grandmother Rachell Day, and her twin brother John M. Day. They were born in Northumberland County, Virginia. Their mother, Mary Day, and their father Daniel Webb, were also located.

Included in this report is information on DNA, the latest technology in locating ancestors. The DNA and genealogical connection is outlined for those who may be interested in learning more about this type of research. Also included:

- The award-winning 'Tracing My African American Ancestors.'
- Review of 15 research steps and other tips.
- Resources that were used to trace the Day family are reviewed.
- There is an extensive review of books that deal with free African Americans.
- Included, is a section on the link between DNA and Genealogical Research, and how it is used to further personal, family, cultural, and historical education. There is also a section on Africa and the connection to America.
- Through DNA testing, two families discover their connections to the KRU people of Liberia, the IBO people of Nigeria, and the Hausa people of Nigeria.

Aaron L. Day, January, 2011

THE DAY FAMILY: A CHRONOLOGY--------------------(Paternal family of Aaron L. Day)

Generation 1   -   1671 -   Mary Day-----Daniel Webb – 1666 - (no marriage record)

Generation 2   -   1692 -   Rachell Day (Webb) ----- Arthur/Will – (no marriage records)

Generation 3   -   1712 -   Winnifred Day (Webb)

Generation 4   -   1735 -   Thomas Day, I

Generation 5   -   1760 -   Rachel Day

Generation 6   -   1777 -   Thomas R. Day, II --- Millie – (no marriage record)-CENSUS

Generation 7   -   1803   -   Scott Day ----- Setta

Generation 8   -   1843 -   Milly Day ----- Washington Day

Generation 9   -   1862 -   James L. Day ----- Sallie Bumpas

Generation 10 -   1894 -   Wilborn Day ----- Ila Torain
**   (Wilborn & James - Brothers)**
Generation 10 -   1907 -   James H. Day ----- Helen/Lula/Lena

Generation 11 -   1935 -   Dora Day-Parks ----- James Parks
**   (Dora & Aaron - Cousins)**
Generation 11 -   1939 -   Aaron L. Day ----- Dorothy Lindsay

Generation 12 -   1959 -   James Parks, Jr.
Generation 12 -   1960 -   Kevin Parks
Generation 12 -   1961 -   Carla Day-Moody-Harris ----- Ray/Jerry

Generation 13 -   1980 -   Jennifer Parks
Generation 13 -   1981 -   Terrell Parks
Generation 13 -   1982 -   Darnell Parks
Generation 13 -   1983 -   Crystal Moody ----- Sean Anthony Harris
Generation 13 -   1985 -   Ray Moody, Jr.
Generation 13 -   1989 -   Ephrom Moody
Generation 13 -   1993 -   Arrionne Moody
Generation 13 -   1999 -   Maliah Harris

Generation 14 -   2000 -   Jahler Parks
Generation 14 -   2003 -   A'Sean D'Anthony Harris - (Crystal's sons)
Generation 14 -   2007 -   Tyray Harris

THE BANKS FAMILY: A CHRONOLOGY----------------(Maternal family of Aaron L. Day)

Generation 1 - Oscar Daniel Banks --------Caroline Penn

Generation 2 - George Banks --------------Eliza

Generation 3 - George H. Banks----- Olla Bondurant - Banks

Generation 4 - Emma Banks-Thompson-Thomas----- James/Chris
(Emma & Lula - Sisters)
Generation 4 - Lula Banks-Day--------James H. Day

OLDEST LIVING GENERATION

Generation 5 - Cordelia Thompson-Hill--------Richard Hill
Betty Thompson-Cox----------Charles Cox
Carol Jean Thompson-Parker (deceased) -------Curtis Parker
Lucy Ann Thompson- ----------------Gordon Austin/Ron Johnson

(Cousins)

Generation 5 -   William Day ------------- Beverly Pullen/Kathy Jones
Shirley Ann Day - deceased in childhood
James Day, Jr., - (deceased) ------ Shirley Freeman/Wadie Meaux
***   Aaron L. Day -   -------------------- Dorothy Lindsay**
Theodore Day -------------------------Carolyn Fairfax
Frederick Day ----------------------Rosemary James
Sharon Lou Day ------------------Doug Thompson

Generation 6 - *Carla Day-Moody-Harris ----- Ray Moody/Jerry Harris

Generation 7 -   Crystal Moody ----- Sean Anthony Harris
Generation 7 -   Ray Moody, Jr.
Generation 7 -   Ephrom Moody
Generation 7 -   Arrionne Moody
Generation 7 -   Maliah Harris

Generation 8 -   A'Sean D'Anthony Harris - (Crystal's sons)
Generation 8 -   Tyray Harris

FOUR- GENERATION 'DAY' FAMILY PICTURE
Front Row, L-R, Maliah, Ray, Jr. (3), Tyray (4)-(Inset)
M. Row, Ephrom, Arrionne, Crystal (3), husband Sean Anthony, & son A'Sean D'Anthony (4)
Back Row, Carla (2), Aaron (1), Jerry- (Carla's husband)

CARLA

Carla was a beautiful child
Whose enthusiasm was something to see
As she grew up.

Carla was a beautiful girl
With a smile that drove
The young boys' wild.

Carla was a beautiful young woman
Who enjoyed living life
To the fullest.

Carla is now a beautiful wife and mother
With five wonderful children
And an adoring husband.

Carla is my beautiful and caring daughter
Who I love very much, and her smile
Still melts my heart.

By Aaron L. Day-For my daughter

FOUR- GENERATION 'BANKS' FAMILY Picture

(LR)   Aunt Emma (3), and daughters; Carol, Betty, Cordelia (4), (seated-George Banks - #1), behind him is his son, George H. Banks (2)

DEAR AUNT

Did you know, how very proud I was when you were named 'Citizen of the Year' for your work within the community? Your many years of volunteer work for the community inspired me very much.

Did you know, that because of your numerous volunteer activities, how many lives would be affected? Through your countless number of dedicated hours, an adult education program was implemented.

Did you know, what a tremendous impact our last conversation had on my life and me? We spent an hour or so, talking, as I learned more about your life, and welcomed the advice you gave me.

Did you know, during that last hour we spent together, that I learned so much about giving and volunteering? I remember, you told me – oh, I just wish we had more time to talk this evening!

Did you know, that I also wanted to do something within the community, and to give back? But it was not until many years later, that I would realize what it was that I had to give back.

Did you know, that I felt as if you had passed the torch on to me to continue your life-long work? Surely you must know by now, how very much I have been enjoying my volunteer work?

Did you know, when I received the 'Volunteer of the Year Award' for my work within the community that I was thinking of you? How very proud and fortunate I am, to have known you.

By Aaron L. Day,
For My Aunt Emma

CHAPTER 1

TRACING MY

AFRICAN AMERICAN ANCESTORS

'Tracing My African American Ancestors' was submitted to the Southern California Genealogical Society for its First Annual National Writing Contest for 2000. It was entered in Family/Local History, or Biological Sketches (Category 1), and was awarded second place. 'Tracing My African American Ancestors' is an excerpt from the following extended 'Day' family history story.

Tracing My African American Ancestors

I remember growing up in a small town called Xenia, Ohio, about twenty miles Northeast of Dayton. There were restaurants where we could not eat. There were neighborhoods we were able to work in, but could not live in. I was five years old in 1944 when my family moved to 421 East Market Street. That home was one block away from what we referred to as the dividing line. Whites lived in one neighborhood, and Negroes lived in the other.

In the neighboring City of Springfield, Ohio, there were certain movie theatres that we were not allowed to attend. I learned about this at the age of 16, soon after I received my first driver's license. I drove into Springfield one evening expecting to see a new movie, and was very disappointed. The cashier told me that the theatre was "by invitation only," in other words, Negroes were not allowed.

The schools were also segregated during the years that I was attending. I was fourteen years old in 1954 when the 'Brown Vs. Board Of Education' decision was passed. It was the landmark Supreme Court decision that declared, 'separate education facilities are inherently unequal.' I will always remember the excitement that was created in my community, and I was especially happy when I heard of the decision to build a new, larger, and integrated high school in Xenia. I could also tell that both of my parents were very pleased with the idea of a new school that their children would be able to attend. Education was very important to them, especially my mother.

My mother's parents had moved from Kentucky to Wilberforce, Ohio so that their two children could attend the schools there. My mother graduated from Wilberforce University in 1931. Wilberforce University was founded in 1865 by members of the Methodist Episcopal church, and is the oldest historically black private college in the United States.

My parents knew how excited I was at being able to attend my last year in a newly integrated high school. You can imagine my disappointment when I learned that the construction of the new school was behind schedule by one year. In 1957, I was in the last graduating class of the segregated East High School. The following year all students, Negroes and Whites attended the new integrated Central High School.

I still maintain close contact with one of the teachers who helped with the School Yearbook. Her name is Edith Keith Jones, and she lives in Irvine, California, not too far from Long Beach, California where I live. I had a great time during my Senior year at East High School. I was the co-editor of our School Year Book, and it was fun putting it together. During my Junior and Senior years, I was editor of the School Newspaper, so I have some very wonderful memories.

In 1964, the landmark Civil Rights Act was passed. It was 'United States Legislation intended to end discrimination based on race, religion, or national origin.' Also in 1964, I graduated from the Dayton Technical Institute, and was the only Black in the class. I will always remember being told by one of the students "I don't know why you are bothering to take this class, you know you are not going to be able to get a job." I was so very proud of my classmates who came to my defense. They told me not to pay any attention to him, because he had problems. They also let him know that he was never to say anything like that to me again. His words were very challenging for me because as a Black person, I knew that it would be extremely difficult to get a job within that field. However, after moving to Los Angeles, California later that year, I was able to eventually find a job within that field. I became one of the first Black Instructors at a Mail-Order Company, and began training others to operate the machines that soon became very popular, and are now known as computers. Later, I received degrees in Accounting and Business Administration.

As a result of the Civil Rights Movement, those of us who lived through that period of time got caught up in the struggle. It was impossible not to be affected, because the movement spread throughout the United States. This was a very difficult period for the United States of America. There were marches, riots, killings, sit-ins, and demonstrations. It turned out that many of us became unofficial pioneers and ambassadors, pushing more and more for our civil rights. We began living, eating, working, attending schools, and going where no Blacks had ever been allowed. There were numerous times that I discovered I was the first Black person to ever work within a particular company or department.

I was living in Los Angeles, California during the 1960's and the 1970's, and very fortunate to be working for a large National Bank, one of the leaders in the banking industry. It was there that I had an opportunity to work with people of many different ethnic groups who were from various parts of the world. Due in part to the Civil Rights Movement, many companies, including the bank, began addressing job discrimination problems such as hiring, firing, and promoting. Other issues such as sexual harassment, and women's rights were also addressed during that time. Many companies played favoritism, were unfair with pay raises, often refused to promote qualified people, and discriminated in other areas. I also faced discrimination in many ways during those years such as racial, job, and housing problems.

In the 1970's, I was asked if I would be willing to relocate to San Diego, California. Administrators at the bank wanted to integrate the Fiduciary Services Department in that city. There had never been a Black person to hold an administrative position in that department. I moved to San Diego that year and joined the Fiduciary Services Department as a Trust Administrator. Later, another Black person joined the department.

In the 1980's, I became the first Black supervisor in the Accounts Receivable Department at a major Electronics Company. There were so many changes going on in the country during that decade. It was that decade that I remember becoming really involved with our family reunions.

The 1982 'Banks' Family Reunion was especially meaningful for me. The Banks family members are very close, and enjoy spending time together. The previous year, my cousin Betty Cox, had collected favorite recipes from all of us, and had a book published, in September 1982. The book is entitled 'Our Culinary Heritage-Favorite Recipes from the Banks family.' In memory of my mother, Lula Banks Day, I contributed three recipes. Betty also included several recipes in memory of her mother, Emma Banks-Thompson-Thomas. My mother Lula, and Emma were sisters.

The cookbook is dedicated to my great-great-grandparents, George and Eliza Banks. The inscription reads; the present members of the Banks family dedicate this book in loving memory of our past generations and leave it as a legacy to the generations yet to come.

We have a Banks family reunion every year. However, since I live in California, and most of the reunions are held in Ohio, or nearby, I am unable to attend every year. Several of my cousins who live in California, and I hosted the reunion in California one year, and a total of eighty-five family members attended. My cousin Olivia Banks, her daughter Susan, our cousin Pam Banks, and I met the weekend before the Reunion and did a trial run of the events.

We planned activities for Friday, Saturday, and Sunday during the family reunion. My youngest brother Fred, cousin Conrad Hill, and his sixteen-year old son stayed at my house in Long Beach, California. I took them to Burbank early Friday morning for the Universal Studios Tour. It seemed like every event we went to, volunteers were being called from the audience to participate, and my cousin Conrad was selected three times. We all enjoyed ourselves. Later that evening, we had a large party in the huge yard at my cousin Olivia's house.

Saturday was a long fun-filled day, beginning early in the morning. All eighty-five of us met at the hotel where the majority of the relatives were staying. We had rented five vans for a tour, and everyone was eager to see the sights. We toured the City of Los Angeles that morning, stopping off at the garment district (the alley), for several hours so everyone could shop. For lunch, we drove into the City of Long Beach and stopped at the Marina, where the Queen Mary ship could be seen in the distance. Later in the afternoon, we drove to Venice Beach for several hours of shopping and sightseeing. Then it was off to Hollywood, and Beverly Hills to see if any movie stars were out.

Later in the evening, it was free time for everyone to do whatever he or she wanted. Some had theatre tickets, movie tickets, others continued to shop, and a few relatives decided to relax at the hotel. I had four tickets for a baseball game that evening in Anaheim, and invited my brother Fred, and several cousins, Carol Parker, and her sister Lucy Thompson. It turned out to be one of the most exciting games I have ever seen.

On Sunday, we reserved a section at the park, and had our annual family meeting and potluck. My cousin Cordelia Hill was the current president of the family association, and plans were made for the next year's reunion. Later in the evening, there was time for more sightseeing. I ended up with a group of relatives' downtown Los Angeles, at the famous Olvera Street area. Everyone enjoyed great meals, and of course, more shopping.

The African-American Civil War Memorial was unveiled and dedicated, July 1998, in Washington, DC. In August of that year, I had an opportunity to attend the Banks' Family Reunion in Ohio. My family dedicated that year's reunion to my great-great grandfather (on my mother's side of the family), **Oscar Daniel Banks.** He was in the Civil War, 1864, and a member of the 100th Regiment, Company I, of the Colored Volunteer Troops. This Volunteer Troop fought for the Union during the Civil War.

The vision for the memorial started with a Civil War Soldier named **George Washington Williams.** He was also an historian and author of Negro History. Williams was the son of a laborer, and he enlisted in the Union Army at 14, and fought in the Civil War. Later, he became a minister, edited and published several journals, and eventually served in the Ohio House of Representatives. He became interested in history, and after doing research, he published 'History of the Negro Race in America."

It was the research for his next book that led to an interest in the establishment of a Civil War Memorial. Williams collected the oral histories of Black Civil War Veterans, and did other research to publish his 'A History of the Negro Troops in the War of the Rebellion.'

On July 2, 1991, **Frank Smith, D.C. Ward 1, Councilmember,** introduced a resolution, which was unanimously passed by the District of Columbia Council. It called for the establishment of the African American Civil War Memorial. This was the first National Monument to honor the U.S. Colored Troops who served in the Civil War. **Sculptor Ed Hamilton**, was commissioned to do the 'Spirit of Freedom' statue.

Sculptor, Ed Hamilton was born on February 14, 1947, in Cincinnati, Ohio. He spent most of his early childhood in Louisville, Kentucky. He graduated in 1969, from the Art Center School in Louisville. After graduation, Hamilton had the opportunity to work and study with the well-known sculptor, Barney Bright. Before being assigned the Spirit of Freedom project, Hamilton created several outstanding works. Hamilton had been given a commission from Hampton University to create a statue of Booker T. Washington. He was later commissioned by the City of Detroit to do a monument honoring heavyweight boxing champion, Joe Lewis. Hamilton also created the famous Amistad Memorial.

The name of Oscar Daniel Banks is on the African American Civil War Memorial. My great-great grandfather Oscar Daniel Banks gave his life in 1864, so that his descendants might have better lives. At the time of his death, the company was posted near Kinsel Springs in Tennessee. This information is based on documents received from the National Archives. The Treasury Department had received an application for his pension, from two of his sons. This document listed the name of Oscars' previous owner. My cousins Cordelia Hill, Charles Banks, and several other relatives did a tremendous amount of research to uncover this information. The following is information by cousin Whitney Anderson, which was passed on at our reunion.

"Based on documents recently received from the National Archives, here is a brief summary of what is known of our history at this time. While several family members had tried to acquire this information a number of years ago, it was difficult to obtain without 'being there.' The advent of computer online services has provided a wealth of information, which has made this search easier. Thanks to cousin Cordelia Hill for obtaining the pension application from which this summary is drawn. We would also like to thank cousin Charles Banks, for following through with a visit to the archives, where he secured further documentation not sent to Cordelia. Charles and cousin Betty Cox, Cordelia's sister, recently attended the Nation's first memorial celebration of 'Colored' Civil War Veterans, held in Washington, D.C.

A note from the Treasury Department, Second Auditors office, Washington, D.C. dated April 21, 1888 reads; In application executed July 15, 1886, George and Parke Banks, minor children of soldier, give their ages as thirty-seven, and twenty-seven years respectively. They state that Oscar Banks and Caroline Penn were married in 1843, at Chas. Given's in Bracken Co., KY., by one Lewis Lightfoot, minister of the Gospel. Upon receiving application for this pension, it was necessary for the Treasury Department to investigate facts establishing legitimacy of the claim. Affidavits were taken from people who knew George and Parke, and or their parents. The list of those giving testimony, along with the Banks brothers included the following.

David Adams (colored)
Lou Clark (colored)
Morgan Davis (colored)
William Gamby (colored)
Phanie Gamby (colored)
Martha Baldwin
William W. Baldwin
Leslie H. Mannen (Banks children were raised on the Mannen family tobacco farm in Mason Co.)
John H. Walton ('owner' of Daniel at the time of his enlistment)"

Whitney did a great job of compiling this information about the Banks family. All families who attended the reunion received copies of the family history. Cousin Cordelia also passed out copies of the Banks family tree.

Oscar Banks had three sons, and two daughters, and I am a descendant of his son, George Banks, who was my great-grandfather. My grandfather was also a George. I loved him dearly, and have very fond memories of him and my grandmother. They lived about five miles from us, and my brothers, sister and I loved to go visit them. Like many grandparents, they had a tendency to spoil their grandchildren, and we loved it.

I am very proud of my great-great grandfather, and was especially thankful for the opportunity to learn of his achievements, at the reunion. Over 100 family members attended our reunion that year. That particular reunion brought us all closer together, and we encouraged families to talk to their children and tell them about their heritage. My oldest grandson Ray was especially proud of his nametag because it had so many greats on it. He is a great-great-great-great grandson of Oscar Daniel Banks.

With all of this information about Oscar Daniel Banks, I knew that I wanted to visit the Civil War Memorial in Washington, D.C. I also wanted my family to share in the experience of visiting the memorial. In 2006, I had an opportunity to take my youngest grandson Ephrom, who had just turned sixteen, to the Million March and More, and he really enjoyed his trip to the Civil War Memorial.

Various   Banks Family Reunions – Ohio

Olla Banks (Mamma Banks), grandmother of Aaron L. Day – Springfield, Ohio Church Convention.   Top - second from right.

Top left - right, Aunt Emma (daughter of Poppa Banks) & husband, Uncle Chris Thomas, George H. Banks (Poppa Banks), grandfather of Aaron L. Day

Bottom - Aaron (Top row, 5th from left)-   Second Grade Class

Day Family & Children – TOP PHOTO, bottom row, L-R, Wadie Day, Beverly Day, Carolyn Day, Rosemary Day, top row, Theodore Day, Aaron L. Day, James Day, Jr., William Day, Doug Thompson, Frederick Day. (Sister Sharon Day-Thompson) took this picture - her husband is Doug Thompson. (Aaron's wife Dorothy is missing from this picture). BOTTOM PHOTO, bottom row, L-R, Kelly Day, Rhonda Day, Pat Day, Carla Day, (baby – Sharon's daughter La Dana), Wade Day, Sean Day, top row, Greg Day, Theodore Day, Jr., Keith Day, Billy Day

In 2006, we celebrated the 25 year anniversary of the Banks family reunions. However, it was after the 1998 Banks Family Reunion that I realized how little I knew about my father's family. Unfortunately, my father had already passed away, and he was the last of that generation.

I began thinking about my father's family, and how little I knew about them. It was very sad for me, because I never met his parents. My father owned a great collection of photos that he had collected through the years. He also had a large collection of letters that were received from family members over the years. I inherited many of these items in 1988. However, it was not until about ten years later, that I really committed myself to searching the Day family heritage.

On September 26, 1998, the Day family members (my father's family) held a reunion. The site of the Day family reunion was the Cleggs Chapel Baptist Church, of Timberlake, North Carolina. After the Reunion, Dora M. Day-Parks (who is a cousin) and I talked about doing a research project for the Day family. After notifying other family members of our idea, we began our research.

People from every period have been curious about their family history, and we are no different. We want to know our family history because learning about our ancestors and their accomplishments will give us a sense of pride; their pioneering efforts have made many things easier for us; and most of all, our children need to know their heritage.

Dora's father, Wilborn Day, and my father James H. Day, were brothers, who both grew up in Person County, North Carolina, near Roxboro. Dora and I have always maintained a good relationship over the years. We have routinely kept each other posted about family members and activities. Dora and I have exchanged Christmas Cards faithfully over the past three decades. I remember summer visits from Dora to our home in Ohio, and the wonderful times we had. She now lives in Baltimore, Maryland, and I live in Long Beach, California.

I began my 'Day' family research at the National Archives and Records Administration in Laguna Niguel, California. In the meantime, my cousin Dora was busy doing research from Baltimore, Maryland. We were on our phones for hours and hours, as we talked about what we wanted to do. We wanted to learn more about where our ancestors were born, and how they lived. What follows, are some of the interesting things Dora and I discovered about our ancestors, and where they lived.

On August 27, 2000, my cousin Dora and I met in Durham, North Carolina for the Day family reunion. Dora, my two brothers, William and Fred, and I drove into Roxboro that afternoon for the reunion. My brother, Ted, and sister, Sharon, were unable to attend. I had arrived from my home in Long Beach, California a week earlier as planned, to do some research.

The week before the reunion, I had driven into Raleigh to do some research at the state library and archives. I discovered a tremendous amount of information about Day ancestors, and the staff members were very helpful to me. The state archives are an excellent place, and I enjoyed the many hours I spent there. I visited the libraries of

Raleigh, Durham, and Roxboro during my trip. I was able to verify some of the information I had accumulated over the years.

Our family reunion was held in Roxboro, which is about twenty-eight miles northwest of Durham. Raleigh is located about twenty-one miles southeast of Durham, and I did a great deal of research there. Staying in Durham worked out great for me because it was centrally located, and it was very convenient for me to move about from one area to another in my rental car.

This was my first visit to North Carolina, and I was also anxious to do some sightseeing. I spent my first day in Raleigh, exploring, three free state museums - Art, History and Natural Sciences - the North Carolina State Archives, the State Capitol building, the Legislative building, and the Capital area Visitor Center. I marveled at the sight of the beautiful well preserved historic buildings.

When I arrived at the Visitor Center the receptionist told me I was just in time to see a short film about the history of Raleigh, the capital of North Carolina. I viewed the film with other tourists, and it gave me an idea of what I wanted to see during my visit. I fell in love with the beautiful city of Raleigh. The city showcases the state's history, culture, the arts, and natural beauty. I experienced first-hand the area's famed southern hospitality. The people were warm, gracious, and very helpful to the other tourists, and me. The tourists I met during my visit were equally fascinated, with this beautiful city.

One of my greatest thrills came that day as I was visiting the North Carolina Museum of History. I discovered a statue and a picture of the famed Thomas Day in front of the museum. Thomas Day was a 19th-century African-American cabinetmaker. Some of his works were included in the exhibit "The Past in Progress" at the museum.

The statue of Thomas was one of three that is located at the entrance to the Museum. There is one of Colonel Fred A. Olds who is the founder, and 'Father' of the North Carolina Museum of History. He traveled to every county looking for objects and their stories to teach the public about the state's past. The other statue is of 'The Sauratown Woman,' a member of the Saura Indian Tribe, who lived along the banks of the Dan River in North Carolina in the late 1600's.

During my research to see if George or Thomas was my great-great-great-grandfather, I had located Thomas Day (the cabinetmaker) who lived in Caswell County, North Carolina. I originally mistook him for the Thomas that I was looking for, who was from Person County, North Carolina. Through court records, I found that the mothers of George and Thomas, (Ann and Rachel) were possibly the sisters or cousins of a John Day. This John Day and his wife Mourning Stewart, were the parents of Thomas Day, the Free Cabinetmaker.

The fact that my Day ancestors were free has made it somewhat easier for me to tract them, through the many different types of documents that are available. There are register records, land deeds, tax records, the U.S. census reports, court records, and other documents that have left a paper trail for those who are tracing free African-American ancestors.

For those who have discovered that their ancestors were slaves, as was the case of my great-great-grandfather Oscar Daniel Banks, there are still documents available. The approach in tracing slave ancestors is somewhat different, and it is very time consuming, but there are some records that will aid in the research. These include; wills, estate inventories, court records, slave schedules, tax lists, bills of sale, Freedman's Bureau records, deeds, and other documents.

Over the previous years, I have read several books and numerous articles about Thomas Day, the cabinetmaker. Thomas was a skilled craftsman of Milton, N.C., and taught his apprentices the art of cabinetmaking. The Yellow Tavern, (also known as Union Tavern) was known not as a Tavern but as the workshop of Thomas. He was one of the great black artisans and furniture makers of the Deep South prior to the Civil War. He trained the slaves of wealthy whites and employed white apprentices to assist him.

Black furniture-makers were the master craftsmen among carpenters of that era. Thomas manufactured fine furniture of mahogany, walnut, rosewood, and cherry. He was known for his staircases, newel posts, and mantels. The very unusual patterns and motifs that Thomas used to highlight numerous stairways and architectural interiors made his work highly popular.

Thomas achieved a fine reputation for himself in Milton. He designed fine large-scale empire style furniture with individualistic decoration and interior architectural trim. The Department of the Interior on May 15, 1975 declared the Yellow Tavern a National Historic Landmark. Examples of Thomas Day's furniture can be seen in the North Carolina State Museum. The curator of the museum told me, they now have a total of 44 Thomas Day items.

Thomas made a deal with the members of the, Milton Presbyterian Church. He donated pews of walnut, yellow poplar, and pine with gracefully curved arms to the church. In exchange, he and his family were given the privilege of sitting in the main area of the church. This area of the church was usually reserved for whites only. Blacks sat in the Gallery.

I was unable to stop by the church to view the pews during this visit. There was so much happening on the day that I was scheduled to visit the church. This will surely be added to my list of things to do on my next trip. Now that I am familiar with the area, I will be able to plan a better itinerary for my next vacation to North Carolina.

The following day, I drove to Person County, North Carolina, and the birthplace of my father, James Day. I met several relatives, and we took a tour of Roxboro. Later in the afternoon I drove into the town of Milton, the home of Thomas Day. It was a great experience to finally view the famous "Yellow Tavern." I had seen many pictures and heard numerous stories about this famous National Historical Landmark.

A major fire had damaged the building, and restoration was in progress. The Thomas Day House/Union Tavern Restoration, Inc. had just received a $250,000 matching Federal Grant for the restoration. Marian Thomas, president of the non-profit corporation restoring Tom Day's home and shop was very happy to receive the grant. She said "winning the

Federal 'Save America's Treasures' Grant underscores the historical significance of the project, and the positive effect its restoration is having on the town of Milton and all of Caswell County." There were 350 applications submitted for the Federal Grants. The Tom Day project was one of only 47 chosen nationwide - and was the only one picked in the state. Marian Thomas said, "Her group willingly accepts the challenge to raise $250,000 to match the Federal Grant."

There were several men working on the Tavern when I arrived, and they showed me around. I was able to get some great pictures of the Tavern and the surrounding area. It is good to know that this historical landmark will be preserved for future generations, thanks to the many dedicated people of Caswell County. I talked to Marian Thomas about the project, and told her how wonderful I thought it was. I joined the Caswell Historical Society, and began actively supporting the group.

Approximately two hundred people attended our family reunion on Sunday. I was able to meet a lot of relatives, some I have known for years, and others that I met that day. The entertainment was great, and the food was delicious. We passed a sign-in sheet around for names, addresses, and telephone numbers, which I added to my computer database. A number of the younger generations were very interested in the family charts that I showed, and wanted to learn more about genealogy, and our ancestors.

During the coming years, I will keep in touch with these young people, and encourage their interest in genealogy. To spark their curiosity, and maintain that interest, I will send family charts and group sheets to them that include their immediate families, and ask if they will help me to keep the Charts updated. I will also make sure they receive a list of all family members, so they will be able to keep in touch. Finally, I will guide them through the research process, so they will learn first-hand how it is done.

My trip to North Carolina was everything that I had hoped it would be; sightseeing, researching, and meeting family and friends. I had only one disappointment; I was still unable to determine my great-great-grandfather Scott's father - Thomas or George. When I eventually discover his identity, then I will know if Ann or Rachel is my great-great-great-great-grandmother.

Through court records from the state of Virginia, I have been able to locate three generations beyond Ann and Rachel. Hopefully, by our next reunion I will have been able to verify this information so I can present it to the family. My oldest brother William is a great-grandfather himself, so we would then have 13 generations of Days.

The August 1998 Banks family reunion honoring my great-great-grandfather, Oscar Daniel Banks changed my life tremendously. I now realize the importance of preserving family histories for future generations. With my family members, (Banks and Days) I am now doing my part to learn about our ancestors. This information will be passed on to our descendants so that they will know their heritage. Hopefully, you have been inspired, and will want to learn more about your family heritage, so that you may pass valuable information on to your descendants.

CHAPTER 2

AARON L. DAY & FAMILY

GENERATIONS – 11-14

Aaron L. Day & Family – Generations 11-14   (The oldest living generation)

This chapter tells the story of the oldest living generation, grandchildren of James L. Day, my paternal grandfather. The brothers and sisters of this generation will be reviewed. We are now the oldest living generation of the 'Day' family. Some of the members of this generation have children, grandchildren, and even great-grandchildren. This history will be passed on to future generations.

I remember my early History Classes, and enjoyed them, although, I didn't care for all of the homework that was required. Learning about history now, is somewhat different from when I attended school and had to memorize historic situations, names, and dates, which, at times, were not always interesting to me. Even the family stories that I heard from my parents and grandparents were sometimes boring, because I had heard them so many times before. Little did I know that someday, I would be anxiously trying to recall each and every one of those stories? My grandchildren are all learning about their family heritage. I keep my daughter, and other family members updated, by periodically sending reports of my research progress. I don't know if they are bored yet, but I try to keep the stories as interesting as possible. Although, they will also have to memorize historic situations, names, and dates, my grandchildren will know that their family heritage is a part of – The Nation's History.

William A. Day

There are five Day brothers and one sister, and they have always enjoyed spending time together. Through the years, the brothers were referred to as the Day boys and were well known within the community where they grew up. Their baby sister Sharon, always had plenty of protection, and was well taken care by her brothers.

William, the oldest of the Day brothers was born in 1933, the same year that Congress passed the 1st minimum wage law ($0.33 per hour). That was the same year that the Federal Deposit Insurance Corp. was created. William enjoyed playing sports in school, especially basketball. He married his high school girl friend, Beverly Pullen. They had two children, Billy and Kelly. After graduation from East High School in Xenia, Ohio, William joined the Air Force. He served in the Army Reserve and Air Force Reserve beginning in 1952. While in the service, he became a great Chef, and had an opportunity to visit many countries. Later, he worked at General Motors until his retirement in 1996.

After several years of retirement, he became bored, and in 1998, he decided to go into business for himself. He opened a clothing store and a restaurant in Wilberforce, Ohio, near Central State University. Like his father James, Williams' hobbies are also photography and gardening. In 2000, William married Kathy Jones of Springfield, Ohio. William has one granddaughter, and three great-grandchildren.

James G. Day, Jr.

James, was the second oldest of the Day brothers. He was born the same year the Ocean Liner 'Queen Mary' was launched at Glasgow. It is now a tourist attraction in the City of Long Beach, California. James attended Lincoln Grade School in Xenia, Ohio, and graduated from East High School. He served in the Army, and was an employee of Dayton Press in Dayton, Ohio.

James attended church regularly and became a member of the East Main Street Christian Church at an early age. He enjoyed cooking and began baking pastries when he was in his teens. James was well-known for his chicken pies, and his pound cakes. He was a Chef at Central State University in Wilberforce, Ohio for many years. James was married twice. He married Shirley Freeman, and later he married Wadie Meaux. James became the father of thirteen children. Sadly, James was killed in an automobile accident in December of 1988.

Aaron L Day

Aaron, the middle of the Day brothers, was born in 1939 in Xenia, Ohio. This was the same year the Daughters of the American Revolution refused to let Opera Star, Marian Anderson, sing at Constitution Hall, because she was Black. In High School, Aaron was the Editor of the school newspaper during his Junior and Senior years. He was co-editor of the Senior Yearbook in 1957. He graduated from East High School where he majored in business. In 1964, Aaron graduated from the Dayton Technical Institute. He moved to the Western part of the country in the mid 1960's where he obtained degrees in accounting and business administration.

Aaron lived in Los Angeles, California for approximately 15 years, then moved to San Diego for one year. He eventually moved to the City of Long Beach, California in the early 1980's where he currently resides. Aaron married Dorothy Lindsay of Pulaski, Tennessee, and they have one daughter, Carla Day-Moody-Harris. Carla has five children, two sons, and three daughters. Carla's oldest daughter, Crystal, has two sons. Aaron also enjoys cooking and baking and was a Chef before moving to California. Like his mother Lula, Aaron has always enjoyed reading and writing.

Theodore A. Day

Theodore was born in Xenia, Ohio, and is next to the youngest Day brother. There was a tremendous amount of conflict the year before Theodore was born. That was the year Pearl Harbor was attacked, and Congress declared War on Japan. Several days later, (12-11-41) the U.S. declared War against Germany and Italy. Theodore Day was born in 1942, and went to Lincoln Elementary School. He excelled in the game of basketball, and was always one of the high scorers in the games. He played four years on the Varsity team - skipping the Reserve team. While playing at East High School, he set a scoring record in one game that was never broken in the history of the school.

Theodore attended Tennessee State and also Central State College, where he graduated. He is active in the community and continues to mentor young people in the game of basketball. He enjoys cooking, and usually cooks for the Banks Family Reunions. Theodore married Carolyn Fairfax, and they have two sons, and seven grandchildren.

Frederick H. Day

Frederick was born in Xenia, Ohio, and is the youngest of the Day brothers. The year Fred was born (1944), was the year that President Roosevelt won his 4th term of office. Fred attended Lincoln Grade School, and later, Xenia High School. Fred has been a member of the East Main Street Christian Church since his early childhood. Like all of his other brothers, Fred enjoys cooking, especially baking.

Inspired by his father James, Fred took an early interest in preserving the family history. He has a great collection of family photos. Fred is a past president of the Banks Family Reunions (which is his mother's side of the family). At all of the family reunions, he collects names, addresses, telephone numbers, and email addresses from other family members, and then passes this information on, so everyone will be able to keep in touch. He has also been actively involved as an historian for the 'Day' descendants, (which is his father's side of the family). Fred married Rosemary James of Youngstown, Ohio.

Sharon L. Day

Sharon was born in Springfield, Ohio, September 11, 1955. As the only girl in the family, Sharon enjoyed being the center of attention. Sharon was born the same year that Rosa Parks was arrested for refusing to give up her seat on a bus for a White man, triggering the birth of the Civil Rights Movement. Several days later, on 12-05-95, the start of the year-long boycott of city bussing Montgomery, Alabama began. Sharon graduated from Xenia High School. She still lives in Xenia, Ohio and has been a Nurse's Aide for many years.

Sharon was a fast learner, and forever inquisitive about things. She was always concerned for the well-being of others, which is probably why she chose the field of health care as her career. We all love Sharon, and she has been a welcome addition to our family. In later years, Sharon was a great source of comfort for our mother, as they were always very close. Sharon and her husband Doug Thompson lived with our mother for several years after their children were born. Sharon has a son, and a daughter.

Our parents were very proud of their children, and worked very hard to give us good lives. They were also excellent role models, and good people. We were very fortunate to have had them as our parents.

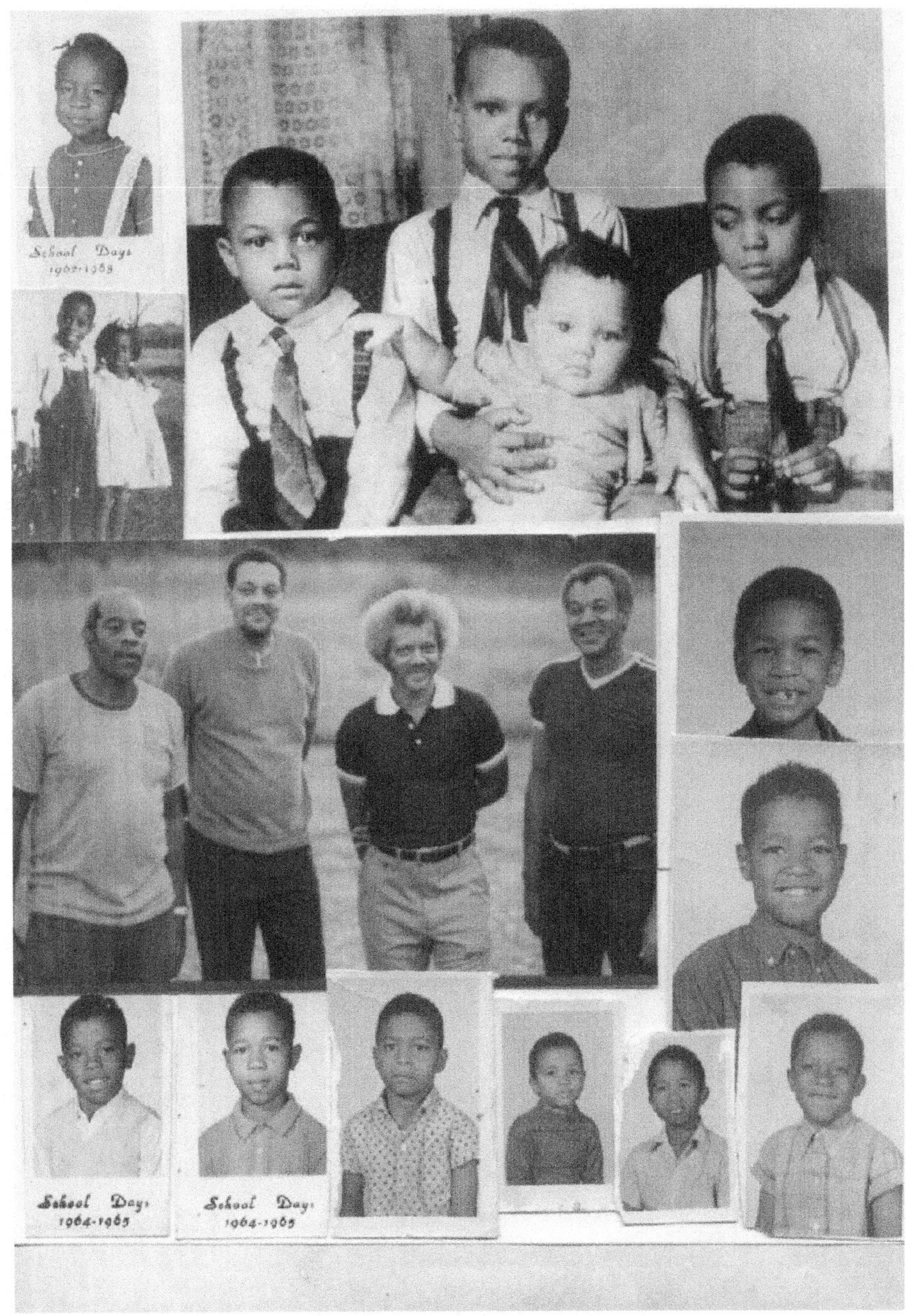

Day Brothers & Sisters

Top left, Sharon Lou Day, William & Shirley Ann Day, Top right, L-R - four of the Day brothers, Theodore, James Jr., Frederick, Aaron

Middle photo, L-R, four of the Day brothers, William, Fred, James Jr., &Theodore – (surrounded by various nephews).

Wedding of Aaron L. Day

Rachel Tilford-Scott, Theodore Day, Mary Adams, William Day, Dorothy Lindsay-Day, Aaron L. Day, Gary Harris, Frederick Day, Wadie Meaux-Day, W. D. Scott. Standing in front, (Sharon Day, sister of the Day brothers)

Various   Day Family Reunions - North Carolina

CHAPTER 3

JAMES H. DAY – GENERATION 10

&

JAMES L. DAY – GENERATION 9

James H. Day - Generation 10

This generation focuses on the lives of the sons of James L. and Sallie Day, my grandparents. They had a total of ten children, seven daughters, and three sons. Two of their sons, (Wilborn and James H.) were Dora's and my fathers.

Wilborn Day, who was the father of Dora, had four sons, and two daughters. He married Ila Torian, and they continued to live in Person County, North Carolina. James H. Day, who was my father, married three times and had five sons and two daughters. James moved to Xenia, Ohio where he raised his family, and spent the rest of his life.

Wilborn and James grew up surrounded by sisters. Wilborn was born in April, 1894, and he was 13 years old when his younger brother James was born. He had been the only boy among five sisters for so many years, he was probably very happy to have a brother. Although there was a 13-year age difference between Wilborn and James, they were always very close. Dora and I often share some of the stories our fathers told us about their childhood days. James always looked up to his older brother Wilborn. Dora tells me that her father Wilborn, said he was always very protective of his younger brother and sisters.

Dora's parents had four sons and two daughters. They are Dallas, Dennis, James, Dora, Annie, and Landers, all born in Person County, North Carolina. Dora Day-Parks was also born in Person County.

James Henry Day, the brother of Wilborn Day was born in Person County on April 10, 1907. James H. Day was the ninth child, and the third son born to James and Sallie Day. The year that James was born, was the year after the Wright brothers of Dayton, Ohio received a patent for their plane. James moved to Xenia, Ohio, about 20 miles Northeast of Dayton, and raised his family there. He was married three times. James lived most of his adult life there, and had a total of seven children.

James married his first wife Helen Greaves, in February, 1929. He and Helen had one daughter: Shirley, and one son, William. Shirley Ann Day was born in Xenia, Ohio, and died at an early age.

In March, 1936, James married his second wife, Lula Mary Banks. They married in Wilberforce, Ohio. Lula was born in Kentucky, April 19, 1905. She was the oldest daughter of George H. Banks, and Olla Bondurant-Banks.

Lula was the first of two daughters born to George, and Olla Banks. Lula was born four days after Andrew Carnegie established the Carnegie Foundation for the advancement of teaching. The Banks family moved to Clinton County, Ohio when Lula was a child. After attending schools there, the family moved once again to Wilberforce, Ohio so that their two daughters might get an education there. Lula graduated from Wilberforce University in 1931.

A religious and faithful person, she was a member of the East Main Street Christian Church, and worked in the Sunday School, the Christian Women's Fellowship, and was a charter member of the Co-Workers Club. Lula died on February 20, 1977, in Xenia, Ohio at the age of 71. James Henry Day, and Lula Mary Banks Day, had four sons, and one daughter: James, Aaron, Theodore, Frederick, and Sharon.

James Henry Day, married his third wife, Lena Stanford in Xenia, Ohio in April, 1967. They had no children together. James was a great communicator, and enjoyed talking to people. He also enjoyed walking, and considered this to be an excellent form of exercise. Despite heart problems through the years, he remained in great condition and continued to take walks. He was a member of the Middle Run Baptist Church. James was a retired Security Guard from Central State University in Wilberforce, Ohio. He had an extensive collection of family photos, and he loved gardening.

My father, James H. Day, was located in the 1920 U.S. Census at age eleven. He was with his mother, father, and four sisters. His father, James L. Day was 59 years old, and married to a Sally L. Day, who was 58 years old. In the occupation column, the census shows that James was the owner of his farm, and the four older children were listed as farm laborers. It further shows that all five of the children were attending school. On the census, Mary is 17, Ollie is 15, Hattie is 13, Jimmie is 11, and Lillie is nine years old.

In 1910, the census reports James at age 4 (probably 4 months), and there is an additional older brother, and an older sister included on this census. Dora's father, Wilborn is the older brother on this census; he is 13 years old. The older sister is Mariah, who is 12 years old. Mary, Ollie, Hattie, and James are also included in this census. The ages on these reports are sometimes inaccurate. The reports are every ten years, and my father was listed as four in 1910, and eleven in 1920. His sister Mary was listed as being one, eleven, and seventeen in three different census schedules.

James is not on the 1900 census, but Wilborn, Mariah, and Mary are included. An older sister named Correnna, who is ten years old appears on this census. From the 1900 census, I learned that James and Sally had been married for 14 years. Column 11 asks: mother of how many children? There are six children listed for Sally. Column 12 asks number of these children living? There are four listed for Sally. I determined that between 1880 and 1900, Sally and James had lost two of their children.

Approximately ninety-nine percent of the 1890 census was destroyed in a major fire, so those records are not available. This is very unfortunate for researchers, because 1890 contained many years of valuable information. I was therefore, unable to find the two deceased children of Sally and James. The years 1880 to 1890 can sometimes be reconstructed through other documents that are available, such as directories, land and deed grants, wills, and court records.

I was researching at the National Archives in Laguna Niguel, California, when I found my father for the first time, in the Census Schedules. It was a great feeling for me to find my father at age eleven, in the 1920 U.S. Census Schedule. Most of you, who have discovered a relative in a Census, know very well that special experience, and feeling.

When I heard someone in the room scream, "I've found my grandmother, I've found my grandmother, she's 10 years old," I turned, as did everyone else in the room, and we shared her joy. We were all very happy for her. It reminded me of my last visit to Las Vegas, when someone's machine hit the jackpot. Everyone within hearing distance stopped for a few moments, and celebrated with him - wishing it had been them.

I began to imagine my father as a boy, and what life was like for him. He did not have the opportunity to spend much time in school, but he learned, and knew so much about life. I know I inherited some of his characteristics and traits. Looking back now, I am so glad I had an opportunity to spend so much time with him.

He was a good man, honest, trustworthy, and dependable. Who paved the way, and set an example for him? Was it his mother, father, grandmother, grandfather, aunt, uncle, or a family friend? My father set a good example for my brothers, sister, and me. He taught us, by leading a good life, and requiring the same from us. We have each, tried to live up to his expectations.

When I was in my mid-twenties, I moved from Ohio to the state of California. My father enjoyed writing, and keeping me posted as to what was going on with the family, and friends. I remember him as being very supportive, and always expressing an interest in whatever I was doing. He left me with some great memories.

As I continue researching the Census Reports for my Ancestors, I reflect upon the time spent with my father. I think about how very much I learned from him, and wonder - why I didn't learn more about him, and his childhood.

James L. Day - Generation 9

I located my grandfather James L. Day, in the 1880, the 1910, and the 1920 Census Schedules. James was with his wife Sallie Bumpers (Sally, Bumpas) and children on the 1910 and 1920 Census Schedules.

In 1981, the Historical Society of Person County, North Carolina published an excellent two-volume set of books about the history, traditions, and families of Person County. I discovered these two-volumes one day while I was doing research at the Huntington Beach Public Library in California. I was very fortunate to find a picture and an article about one of my ancestors, Richmond Day, who was born in 1847. This article was submitted by his great-granddaughter, Lala Webb.

Richmond Day was the son of George and Emily Day. Having purchased his home before the birth of his children, Richmond continued to add to and improve that home and purchase additional land as it became available to him. Richmond, was the cousin that my grandfather was living with during the 1880 U.S. Census. James was nineteen years old, and living with Richmond Day, and his family. Evidently, James went to live with his cousin after the death of his son, who died at an early age. My next challenge was finding the parents of James L. Day.

While checking the 1870 U.S. Census Report I found a Milly Day, and her children. She was living with her three children, Aaron, Henry, and Setta. I was able to determine that

Milly Day, was my great-grandmother because of a family report Dora did when she was in College. When Dora compiled this report, her mother was still living, and she obtained an enormous amount of information from her. In a letter to me Dora says, "I am sending you more information for the Family Tree. It's as follows: I do not know the names of James L. Day's father or mother. These are the brothers of James. There is a sister Jennie, also included." Aaron Day, Gus Day, Fred Day, Henry Day, Floyd Day, James Day, and Jennie Day.

I stared at the list sent to me by my cousin Dora of our grandfather and his siblings. It looked like a list of me, and my brothers. My name is Aaron, my older brother was named James Day, my oldest brother is named William Andrew (nicknamed Gus) Day, my younger brother is named Theodore Augustus Day, and my youngest brother is named, Frederick Henry Day. The similarity of the names was unbelievable.

I found Milly again on the 1880 census with five more children named Thomas, Augustin (Gus), Sulu, Jennie, and Charlie. Our grandfather James and his brothers Fred and Floyd were not on either of these two census schedules with Milly. As noted earlier, James was found living with his cousin Richmond Day in 1880. From Dora's letter, I was able to link - Aaron, Gus, Henry, and Jennie to our great-grandmother Milly Day. The Aaron was my namesake. Through oral history, I remember my father telling me that I was named for his Uncle Aaron.

Dora also sent me detailed information about each of them, including their children. This list from Dora included all ten children of James L. Day, and his wife Sally L. Day. The two children missing from the 1900 Census were identified as Luna, and Crawford. Luna died at the age of thirteen, and Crawford died at age three. Their children were:

(1)   Luna was the oldest, and died at the age of 13
(2)   Crawford was the second child, and the first son. He died at an early age
(3)   Corrina was born March, 1890
(4)   Wilborn was born April, 1894
(5)   Mariah was born October 1896
(6)   Mary was born May, 1899
(7)   Ollie was born in 1902
(8)   Hattie was born April 4, 1904
(9)   James Henry, my father, was born April 10, 1907
(10) Lillie was born in 1910

Top – Left, L.R. Day brothers and cousin Christine Jefferson. James, Christine, Aaron. Bottom – L.R. Theodore, Frederick. Right – Dorothy and Aaron L. Day, with their daughter Carla.

Bottom–Left, Carla, Aaron L. Day. Right–Lena Day, James H. Day, (father of Aaron L. Day) Carla Day, Aaron L. Day

James Ware, World War II Veteran, and nephew of James H. Day.

Helen Greaves-Day, first wife of James H. Day

Top L-R, Unknown Couple on left, Lena Day (3rd wife of James H. Day), and James H. Day. Top far right, William Day & family visiting Uncle Wilborn, (brother of James H. Day) and Aunt Ila Day, in North Carolina.

Middle L-R, Beverly Pullen-Day, Ila, Wilborn. Front, Kelly & Billy Day. Far right James H. Day

Bottom L-R, William Day & Uncle Wilborn Day. Far right Billy Day, Uncle Wilborn, and Kelly Day

Top, L-R, Aunt Ollie Day (sister of James H. Day), and Frederick Day

Right photo far right, Lula Day (2nd wife of James H. Day),   at Church Baby Contest

Middle Photo, left - James H. Day, right Photo – (three generations) Joe Banks, brother of George, Lula Day, Aaron L. Day, grandfather George H. Banks (poppa Banks)

Bottom Left Photo, L-R, bottom row, Rosemary Day, Wadie Day, Cordelia Hill, holding son Ricky Hill, Lucy Thompson, back row, William Day, Frederick Day, Sharon Lou Day-Thompson, Aaron L. Day, Carole Hill, son & husband)

James Day, Jr., & Aaron L. Day, William Day, James Day, Jr., & Shirley Ann Day

**Day brothers & Thompson sisters (cousins) - L-R, Top-Front Row
(All four daughters of Emma Banks-Thompson) Betty Thompson Cox,
Carol Thompson-Parker, Lucy Thompson-Austin-Johnson, L-R, Back Row,
Cordelia Thompson-Hill, Frederick Day, William Day**

**L-R, Middle Row, Cordelia, Betty, Carol, Lucy - L-R, Bottom, – Aaron L. Day, Betty
Cox, William Day, Lucy Johnson, Cordelia Hill, Carol Parker,   (Top) Frederick Day,
and Theodore Day**

George and Eliza Banks

Maternal Great-Grandparents of Aaron L. Day

Betty Cox, had collected favorite recipes from many family members, and had a book published, in September 1982. The book is entitled 'Our Culinary Heritage-Favorite Recipes from the Banks family.' In memory of my mother, Lula Banks Day, I contributed three recipes. Betty also included several recipes in memory of her mother, Emma Banks-Thompson-Thomas. My mother Lula, and Emma were sisters.

Day Family children at Family Reunions – Ohio

CHAPTER 4

MILLY DAY

GENERATION 8

Milly Day - Generation 8

My great-Grandmother Milly Day was born in Person County, North Carolina in 1843. This was the same year that Isabella Van Wagener, Black American Evangelist and Reformer took the name of Sojourner Truth. Traveling up and down the land, Sojourner sang, preached, and debated at camp meetings, churches, and on the streets. As an Evangelist and Reformer, she carried the biblical message of God's goodness and the brotherhood of man. She eventually became an Abolitionist and also supported the Women's Rights Movement.

In 1847, when Milly was four years old, Frederick Douglass began publication of the North Star, an antislavery newspaper. This contributed to his break with White Abolitionist leader William Lloyd Garrison, publisher of the Liberator. Douglass had worked with Garrison in publishing the Liberator. Frederick Douglass was one of the most eminent human-rights leaders of the 19th century. His oratorical and literary brilliance thrust him into the forefront of the U.S. Abolition movement. He became the first Black citizen to hold high rank in the U.S. Government.

Milly was thirteen years old when members of the Methodist Episcopal Church founded "Wilberforce University" in 1856. Wilberforce is the oldest historically Black Private College in the United States. My mother, Lula Banks Day, graduated from Wilberforce University in 1931. After the University was closed during the Civil War, it was bought and reopened by the African Methodist Episcopal Church. I grew up, and lived about four miles from Wilberforce University.

Census Schedules provided information about Milly's children. Milly's first son Aaron was born in 1859 when she was sixteen years old. Aaron appears for the first time on the 1870 census with his mother. The 1880 census indicates that Aaron was working as a hired servant. He was married to a Caroline, age twenty-two, and they were the parents of a seven-month old boy named William.

From oral family history, I can still remember hearing my father talk about the Uncle Aaron that I was named after. Aaron would have been about 48 years old when my father was born in 1907. My father was apparently very close to his Uncle Aaron. He would often tell me when I was doing something, that I reminded him so much of his Uncle. I also remember my father talking about some of his other Uncles and Aunts.

Milly's second son, James L. Day, was Dora's and my grandfather. He was born in December 1862, in Person County, North Carolina. When Dora and I began the Day Family Research Project, it was to learn more about our grandfather, James L. Day. We discovered the year that he was born was a year full of dramatic changes within the country. That year, paper money was issued for the first time in the United States, the U.S. Government began taxing tobacco, and the Civil War continued.

The following year, the Emancipation Proclamation which went into effect on January 1, 1863, was an important moment in the relationship of Frederick Douglass and President Abraham Lincoln. Douglass and Lincoln had tremendous respect for one another.

Later in the year, on Tuesday, September 22nd, 1863, President Lincoln made his Emancipation Proclamation speech. Still later, on November 19, 1863, Lincoln made his celebrated Gettysburg Address. Fourscore-and seven years ago our fathers brought forth on this continent a new nation, conceived in liberty and dedicated to the proposition that all men are created equal.

The Union Army enforced the Emancipation Proclamation. When the slaves realized that they were free, they began to celebrate. Over the years, this has become an annual celebration. It became known as Juneteenth. In 1997, the U.S. Congress recognized June 19, as Juneteenth Independence Day in America.

Although James was born during the early part of the Civil War, it would end in 1865, when he was three years old. It was not until many years later, that he would learn of the turmoil that had taken place during his childhood. In the time of his early childhood, Congress passed the 13th, 14th, and the 15th Amendments.

AMENDMENT XIII

Neither slavery nor involuntary servitude, except as a punishment for crime whereof the party shall have been duly convicted, shall exist within the United States, or any place subject to their jurisdiction. This was signed into law December, 1865.

AMENDMENT XIV

All persons born or naturalized in the United States, and subject to the jurisdiction thereof, are citizens of the United States and of the State wherein they reside. No State shall make or enforce any law which shall abridge the privileges or immunities of citizens of the United States; nor shall any State deprive any person of life, liberty, or property without due process of law; nor deny to any person within its jurisdiction the equal protection of the laws.

AMENDMENT XV

The right of citizens of the United States to vote shall not be denied or abridged by the United States or by any State on account of race, color, or previous condition of servitude.

All three of these amendments had a profound effect on the people of the United States of America, especially the slaves.

There were great social and economic changes during the period immediately following the Civil War. This ten-year period was known as the Reconstruction Era. Congress approved the Freedmen's Bureau on March 3, 1865. This Federal Welfare Agency was created to solve some of the problems the newly freed slaves were facing. Freedmen were able to vote for the first time in the 1867-68 elections. Federal Troops were called in to maintain law and order.

The Freedman's Savings & Trust Company or the Freedman' Bank was chartered on March 3, 1865. It was a completely separate institution from the Freedmen's Bureau. The Freedman's Bank was originally established for Black Union Troops who needed a place to deposit their money. The Bank eventually established thirty-seven branches in 17 States and the District of Columbia.

Milly's third-son Henry was born in 1865. Confederate Generals, Robert E. Lee and J. E. Johnston, surrendered to General Ulysses Grant, and the War was finally over. The North's victory in the American Civil War resulted in the preservation of the Union, the abolition of slavery, and the granting of citizenship to the freed slaves.

Milly's son Henry, appears for the first time on the U.S. Census in 1870 at age five. He and his sister Setta, both appear on the 1880 Census as hired hands, working for a William P. Satterfield. Henry was sixteen years old on this Census. The Satterfield farm was located next to the property of Henry's mother, Milly. William Penn Satterfield had inherited his farm in 1859 upon the death of his father. The Satterfield family dates back to approximately 35 years prior to the founding of Person County.

Since the Census for 1890 was destroyed in a major fire, I went to the 1900 Census in search of Henry and Setta. The 1900 Census is a great source of information for anyone doing family research. It gives so much detailed information about each person and their families, such as; place of birth of each person, as well as the place of birth of their mother and father. It also lists: the number of years in the United States, educational background, and more. I did locate Henry on this Census.

I found Henry at 35 years of age on the 1900 Census, and as Head of Household. The Census revealed that he had been married to Tish for 15 years, and that they had six children. The report further states that all six children were living, and all of them were on this Census. There were five boys and one girl. George was eight, Luther was six, Lambert was four, Owen was three, Millie was two, and Manard was six months. Dora tells me that Henry's great-grandchildren, Linwood, and Michael are now living in Baltimore, Maryland near her.

Setta M. Day, was Milly's first daughter, and fourth child. She was born in 1867. Setta was the first of Milly's children to be born after the Civil War ended. I found Setta on the 1870 Census at three years of age, with her mother Milly, and her two brothers Aaron, and Henry. When I checked the 1880 Census, I found her listed as (Setty). She was with her older brother Henry who was 16 at the time. Setta was 13, and working as a hired hand for William P. Satterfield. I have been unable to find any more information about Setta since the 1880 Census.

Milly's son Thomas M. Day, was born in 1870, when she was twenty-seven years old. This was the year that Hiram R. Revels of Mississippi took the former seat of Jefferson Davis in the U.S. Senate. He became the only Black in the U.S. Congress and the fist elected to the Senate. Also in 1870, Joseph Hayne Rainey was the first Black elected to the U.S. House of Representatives. He was from South Carolina.

Milly's fifth son, and sixth child was named Augustin Day. He was born in Person County, North Carolina in 1872. Augustin eventually moved to Xenia, Ohio where he married Minnie Greaves. Augustin was known affectionately as Uncle Gus. Augustin and Minnie did not have any children, and were very close to their nieces and nephews.

I remember Uncle Gus and Aunt Minnie very well. They lived about five blocks from our house. As a young boy, I remember my father and his Uncle Gus sitting on the porch and talking. They both loved to talk, and would do so for hours, as my brothers and I played games. Aunt Minnie was one of our favorites; she always had good things for us to eat. Aunt Minnie and her sister Barbara Ware, were always baking cookies, pies, cakes and other pastries for us children.

The year that Uncle Gus moved to Ohio is uncertain. It appears he may have had something to do with my father leaving North Carolina, and moving to Ohio. He was the only one of my grandfather's siblings that I ever met.

Sulu Day was born in 1875, in the County of Person, North Carolina. She was Milly's second daughter, and her seventh child. I have been unable to uncover any information about Sulu. The year 1875, was also the year that Mary McLeod Bethune, was born in Mayesville, South Carolina (1875-1955). Mary McLeod Bethune was a great American Educator. On October 3, 1904, Bethune opened Daytona Educational and Industrial School for Negro girls, which later became Bethune College. In 1925, Bethune College and Cookman College merged to form Bethune-Cookman College, in Daytona Beach Florida. In high School, I had to write a paper about a famous Black person, I chose Mary McLeod Bethune.

Jennie Day, was Milly's third daughter, and her eighth child. Jennie was born in Person County, North Carolina in 1878. I have learned that Jennie had two daughters, one named Effie, and one named Sallie.

Throughout the Reconstruction Period, (1865-77) Milly would have been busy raising her children. She would have been 34 years old at the end of Reconstruction in 1877. It was during this time, that Frederick Douglass was fighting for full civil rights for Freedmen. He was also vigorously supporting the Women's Rights Movement. As Milly and her family were fighting for their rights as free people of color, it is likely that they also became actively involved in the Women's Rights Movement.

A review of the 1880 Census shows that Milly was working as a Day Laborer. There is a column for Civil Condition, (Marital Status) on this Census. In column #11, (Widowed/Divorced) there is a check mark for Milly after Widowed. Her husband Washington Day, obviously died sometime within the previous year or so. Milly's youngest son Charlie, is seven months old on this Census. Charlie Day was Milly and Washington's sixth son; he was born in Person County, North Carolina in 1880.

At this time, little is known about Fred and Floyd Day. I have not been able to locate Fred or Floyd on the 1870 or 1880 Census Schedules. I am continuing my research to learn more about their lives. Fred and Floyd were both on the list that Dora prepared for her school project. I am checking the possibility that these were either nicknames or middle names for the other boys.

I was unable to check the 1890 Census because it was destroyed in a fire. Milly was not located on the 1900 Census. I am, however very grateful that we have been able to learn so much about the family of Milly and Washington Day, from the U.S. Census Schedules.

Without the information from these schedules, we would have been extremely limited in our ability to locate many of our ancestors.

Searching for Aaron L. Day

Thank goodness for the Internet, and e-mails. In July of 2006, I received e-mails from two friends, stating that someone was searching for me. The information received, helped me tremendously with my Day family research.

The first e-mail was from my friend Gladys Waters:

Hi Aaron,

Hope all is well? I received the following e-mail from a friend who remembered me mentioning that I had met you.

Take care, Gladys Waters

Hi Gladys,

Nurse Jackson here!!! I received this e-mail and recall you mentioning Aaron Day at one time. Possibly you could forward this attachment to him.

The second e-mail received was through my friend Marjorie Shoales Higgins, President of the California African American Genealogical Society, in Los Angeles.

Hi Aaron,

I thought I remembered your name from the Los Angeles Summit and sent this to Marjorie Higgins. Marjorie gave me your e-mail, and now I'm including a message from someone on a roots web list who is trying to contact you.

MESSAGE: Hi, how would I get in touch with Aaron L. Day? He wrote a book about free African Americans before 1865, and his ancestor was a cabinet maker in Caswell County, North Carolina? I just read an article he wrote about DNA research in Africa. He is also Vice-President for an African American Genealogy Organization in Southern California.

Thanks, Karin Berry,

I immediately contacted Ms. Berry, and she sent me the following e-mail regarding one of my ancestors.

Dear Mr. Day

I just read your article on DNA and researching African ancestry in the July/August 2006 Everton's Genealogical magazine. I have traced my Long and Enoch families to Caswell

County, North Carolina. While exploring the county's Website, I found information about your Day ancestor.

It so happens that Gus (Augustin) Day of Caswell County migrated to Ohio at the same time as my Long family. He was a widower and lived with my great-great Uncle David Long, in Springfield, Ohio. I also found Gus in Caswell when he lived on the same street as my great-great-great-grandfather Nathaniel Long.

Is this your ancestor? I saw the photo in Everton's magazine identifying Augustus (Gus) Day as your grandfather's brother, and his wife.

Thanks for your help,

Karin Berry, in Philadelphia

This was great news for me. I talked to Karin on the telephone, and we decided to share research information. Karin sent me the following e-mail which revealed a great deal of information about my Uncle Gus. I will be forever grateful to her.

HI,

Our family history says that Edward Nathaniel "Pink" Long got into a fight with a White man and had to flee the State, and that the family left soon after. Not sure how accurate this is. I've read that many people migrated in groups. It looks as if Gus either moved with them, or moved soon after they left. I have pinpointed the Longs' departure from North Carolina as between June 1900 (after the Census Enumeration) and December 1900, when their youngest son was born in Ohio.

I found Gus Day in the U.S. Census records, Living near, or with my Enoch family (1900) Series T623, Roll 1180, Page 69.

Gus Day, 24, in Alamance County, North Carolina, Faucette Township, married. He lived next door to Ike Walker, my great-great-grandmother's husband, and their family, wife Sarah, daughters Minnie (my great-grandmother), Bertha, Lila and Bedford. Sarah Walker was the daughter of Bedford and Tenner Enoch, who are recorded on page 70. There were Bedford Enoch, wife Lena, daughter Maggie, and her son Lucian.

The two daughters married two Long brothers; Minnie married Alex Long, and Bertha married David Long.

1910 - Gus, age 33, boarder, living with my great-uncle, David Long, on Miami Street, Springfield, Ohio in 1910. He was widowed. Others in household were David 22, wife Bertha, 20, son George, 5 months. Bertha had two children, only one was alive. They had been married for two years.

1920 - Series T625, roll 1386, Page 268, Gus was living alone in Xenia, Ohio. He was 46 years old, and a laborer at a rope factory.

Karin

This e-mail revealed so much information about my Uncle Gus.

(1) - Now I have an idea why my Uncle Gus moved from North Carolina to Ohio.

(2) - I thought that my father moved to Ohio because of his Uncle Gus, this adds to that theory.

(3) - This information reveals that Uncle Gus was married before moving to Ohio, and had two wives. This accounts for the age differences in the two photos we have of Uncle Gus and his wives. We were under the impression that he had only one wife. On the second picture, he has aged considerably, while she has not aged at all. I did some checking through the City Directories of Xenia, Ohio when I returned for my family reunion, and discovered that he had married Minnie Graves in Xenia, Ohio.

Left - Uncle Augustin (Gus) Day & first wife, Right - Uncle Gus & Aunt Minnie

CHAPTER 5

SCOTT DAY

GENERATION 7

Scott Day - Generation 7, My 2-G-G-Grandfather

The abstract of the Person County, North Carolina Marriage Records, 1792-1868 compiled by noted historian Katherine Kerr Kendall, was the connecting link between Scott, and his daughter Milly. This report contained the following information; On Oct. 25, 1867, there was a Bond for; Washington Day (Groom), Milly Day (Bride) Daughter of Scott and Latta. Bondsman and witness - (Sam'l Y. Brown, Clerk. M. 8 Oct. 1867) by B. A. Thaxton, J. P

Setta's name was written as Latta) on the Marriage Bond, all other spellings have been Setta. Scott and Setta's daughter Milly, (Dora's and my great-grandmother) had a large family. As noted earlier, we verified that she and her husband Washington, had at least, six boys, and three girls. Milly's firstborn was named Aaron, my namesake. They were living in North Carolina.

North Carolina has three distinct and dramatically different land regions; the Atlantic Coastal Plain, the Piedmont Plateau, and the Mountain Region. I began to learn more about the different regions. The Piedmont is the region I found our Day ancestors. They have been inhabitants of this area for many generations. The search for Dora's and my great-great grandfather, Scott Day, started after I located his daughter, Milly on the 1870 U.S. Census Schedule. Scott was with his wife Setta on that Census just above Milly and her family. Scott was 67, and Setta was 65. His occupation was listed as a Hatter. He probably made hats for Person County and the surrounding communities in North Carolina. The Civil War ended in 1865, and everyone was listed on the 1870 U. S. Census, including the newly freed slaves.

On April 12, 1861 When Scott was 58 years old, the Confederates attacked Fort Sumter in South Carolina. This was considered the beginning of the Civil War. President Abraham Lincoln called for 75,000 volunteers to defend the Union. On December 20, of the same year, a pro-slavery convention meeting in South Carolina voted to leave the Union. The American Civil War was fought between 1861 and 1865 and between Northern States who wanted slaves to be free and Southern States supporting slavery.

I have checked the Civil War Soldiers & Sailors System and other Military Records and found no trace of a Scott Day. At 58, it is doubtful that Scott was in the War as a Soldier, although, there were many older men in the War. It must have been very shocking for him and his family to see the terrible effects that the War was having on their community. It is recorded, that civilians in both the North, and the South were working to aid in the War effort, this included free people of color, and slaves. Imagine my surprise when I discovered my great-great-grandfather Scott, on the 1860 U. S. Census Schedule, listed as a free man.

I found Scott listed on the 1860 Census working as a servant for Isaah and Elizabeth Bumpap. On this Census, Elizabeth and Isaah have five daughters, ages eight through twenty-two. It is reported that Isaah is a farmer, and owns his farm. In the description section of the 1860 Census, there are columns for age, sex, and color. In the color column, there are three categories; White, Black, or Mulatto. On this Census, Scott is shown to be a Mulatto. Census Takers counted as Mulattos, only those who appeared to be of mixed parentage (light complexion). In Scott's case, this was the only Census that he was listed as a Mulatto. All other Census Takers had listed Scott as Black. I thought immediately of the book 'Finding a Place Called Home,' by Dee Parmer Woodtor, and the different reason that Scott could have been free during this period. This was the first indication I had that Scott may have been of mixed parentage. I considered this a very important clue in my search for 'Day' ancestors. I grew up hearing stories of the mixed ancestry in my mother's family. Now, I was beginning to learn of the mixed ancestry in my father's family.

Scott appears on the 1850 U.S. Census working as a farmer for 70 year old Mary Williams. At seventy, Mary is Head of Household, and apparently has several daughters living with her. Savannah is 55, and Sarah is 45. Mary also has several other people living with her. Sarah Wyatt, who is a young girl, 12 years old, is a Mulatto. It is unclear what her relationship was to Mary Williams. There is also a man named Charley David, age 22 living on Mary's property, his occupation is given as a Crafter.

When Scott was about 50 years old, William Wells Brown, a former slave, abolitionist, historian, and physician, published Clotel. This was the first novel written by a Negro American, in London. The work is an account of the life of a Black woman whose father was an American President. The book draws on the legend that Thomas Jefferson had fathered many children by his slave mistress. Brown was born to a slave and a White slave-owner in Lexington, Kentucky in 1816. He also published, 'Three Years in Europe: or, 'Places I have Seen and People I have Met.'

When Scott was 44 years old, the courts handed down the Dred Scott decision. Dred Scott was a Black man, once the property of John Emerson, an army doctor who had bought him as a slave to Missouri, and had then taken him into Illinois, a free state, and from there into Wisconsin territory. Dred Scott sued the state courts of Missouri for his freedom, on the grounds that he had lived in free territory. He won his case, but the decision was reversed on appeal.

The year 1849 was also the year that Harriet Tubman, made her daring escape from slavery. During the following decade, she returned to the South many times, helping to guide other slaves to freedom. She became known as the 'Moses' of her people as she helped hundreds of slaves to escape through the Underground Railroad.

Scott Day was also listed on the 1840 Census. There are three sections on the 1840 Census. The first section contains the names of Heads of Families. The heading for the second section reads, Free White Persons, including Heads of Families - with columns for males and females, categorized by ages. The last section titled, Free Colored Persons is where I found Scott. He was under the 24-36 age columns. Our records indicate that Scott would have been 36 or 37 years old in 1840.

The year that Scott turned 24, two Blacks, John Russwurm (the first American Negro College graduate) and Samuel Cornish began publication of Freedom's Journal. This Journal was the nation's pioneer Negro newspaper. They began the paper on March 16, 1827, in New York City, and began speaking out against slavery. Russwurm and Cornish were free long before the Civil War.

By the year 1830, the number of free blacks by the name of Day, in Person County, North Carolina had increased from 16 to 24 on the census. Also, the number of free black households with the name of Day in the State of North Carolina had increased to 15 households, with a total of seventy-one members. A review of the census schedule for 1830, revealed that there were 76 free African American Day households in the United States.

It was around 1832, that a slave rebellion took place. Nat Turner was born a slave on October 2, 1800, in Southampton County, Virginia. His owner's name was Benjamin Turner. Nat began working in the fields when he was twelve years old. Although a slave, Nat learned to read, and his master approved of him reading the bible. Later in his life, Nat became interested in the ministry and began preaching at local churches. He preached to black people about his visions of judgment, and freedom, and had several dreams that he interpreted as a call to freedom. An eclipse of the Sun was a sign to Nat that the time for rebellion had come.

On August 22, 1832, Nat and a group of men met at an old farmhouse, and the rebellion began. Gathering an army of about 60 men, Nat began to move across Southampton County destroying farmhouses along the way, until the militia forced them back. Soon after, the army moved into Southampton County, and the navy followed. There were 57 white people killed in the rebellion.

Shortly afterwards, the rebellion survivors went into hiding, including Nat. After his capture, Nat was questioned, arraigned, and brought to trial. He was eventually hanged. Nat Turner became a symbol of the slaves' determination to win their freedom.

It was very difficult to think about my great-great-grandfather Scott, living under these conditions. To try to better understand this rebellion, and to capture how Scott might have been feeling during this period, I compared the Revolt of 1832, with the Riot of 1965 (The Watts Riot). I was living in the Los Angeles area during the Riot of 1965. Although I lived on the West Side of Los Angeles away from the main area of unrest, the feeling of fear and helplessness spread rapidly throughout the city, the state, and the country.

There were many reasons leading to the Revolt of 1832. Unequal treatment was one of the reasons they were dissatisfied. The slaves were treated as second-class citizens and inferior in every way to the white citizens. The slave worked continuously through the day, from early sunrise to late sunset. Poverty was another reason for the revolt. The majority of the slaves lived a very shabby life in comparison to the white people. Most of the slaves lived in small cabins that were usually very crowded.

As in the Revolt of 1832, the reasons for the Riot of 1965 were many and also very similar. The people of the Watts area were living near poverty levels. Many of the people were living in housing projects, which should have been torn down years before. Some were on welfare, unwed mothers, the hard-core unemployed, and the elderly.

A few of the reasons for the two disturbances were very similar. In the South, after the Civil War, the North had put the South in a very dependent position. During the War, the North passed certain tariffs and resolutions that were to the advantage of the North, and to the disadvantage of the South. As a result, money was flowing out of the community at a faster rate than it was coming in. There was never enough money for improvements within the community.

One similarity I found very interesting was the fact that both the Revolt and the Riot happened in the hot summer month of August. In contrast, the two were very dissimilar as far as leadership. The Revolt of 1832 was a planned and well thought out affair that was under the direct leadership of Nat Turner. In the Watts Riot, there is no evidence that the riot was pre-planned. After the riot started, there may have been some teenagers and young adults who banded together and looted, but this was spontaneous, and not planned ahead. There were mobs breaking into stores, looting and setting fires throughout the Watts neighborhood, and the National Guard was called in to control the unrest. There was a curfew put into place, and people did not want to be out after dark.

Many of the neighborhood food stores were burned out, and residents had to shop during the day at markets outside the area. As a result, it would take hours to buy just a few items such as bread and milk, because of the long lines. Most everyone began filling their shopping carts and stocking up on food, and I found myself doing the same, because it seemed as if the stores might run out of supplies.

There are quite a number of different opinions as to the results of the Riot of 1965. One thing can be said for sure, the riot put Watts in the spotlight. It brought attention to the sad condition of many of the people; the unemployment, the poor housing conditions, the mistreatment by law officials, and the abuse by many of the neighborhood stores, and the high prices for food.

After the Watts Riot, I personally noticed a change in the attitudes of the people who lived in the City of Los Angeles that I came in contact with. There seemed to be a sincere desire among the people to want to get along better, and to try to understand what really happened. Everyone seemed to be in full agreement though, that change must come from outside, and within. There had to be better understanding, caring, and especially a willingness to communicate.

Unlike the Riot of 1965, where there was better understanding and a willingness to communicate afterwards, the Revolt of 1832 took a different turn. The attitudes of Southern whites toward blacks began to harden after the 1832 rebellion led by Nat Turner in Virginia. Fearing a slave revolution, North Carolinians passed harsh laws concerning slaves. Free blacks such as Scott were also affected by the hardening attitudes. After the Nat Turner Rebellion, free blacks were required to wear badges to distinguish them from slaves. Until 1835, North Carolina allowed free blacks to vote. However, the same revised 1835 constitution that gave political power to small farmers, took away voting privileges from free blacks.

The comparison of the Revolt of 1832 with the Riot of 1965 (The Watts Riot), gave me a better understanding as to how my great-great-grandfather Scott may have been feeling when the Revolt took place. During the Riot of 1965, I was very aware of the fear and helplessness all around me. The fight for survival was especially noticeable when I went out shopping for food, and other necessities. There was so much tension, anger, and fear, as people began looking out for their families and loved ones. Yes, it must have been very difficult for Scott and his family.

'Land and Negroes for Sale,' read a sign posted near the Stokes County Courthouse in 1836. The sign also announced the sale of 'corn, oats, and horses,' along with seven Negroes, consisting of men, women, and children.' Such signs were common in the South during this period, when Scott was a young man.

After searching through the entire 1820 census for Person County, North Carolina without finding anyone by the name of Day, I came to the recap-page. You can imagine the sense of disappointment I felt while viewing the recap-page. There were 2,817 white males and 2,615 white females for a total of 5,432. There were also 1,890 male slaves, and 1,804 female slaves listed, for a total of 3,694. I noticed the absence of names for the slaves. There were no slave names given, just numbers listed in columns-for males and females. I imagined that my great-great-grandfather Scott had disappeared into one of these columns, and was now one of the unnamed slaves. My first thought was that I would have to start searching through the Slave Schedules. I realized this would be a challenge, but I knew I had to try to locate my great-great-grandfather Scott.

I did feel very fortunate that I had been able to go back 180 years, and discover so much about my ancestors in such a short period of time. Staff members at the National Archives, and the Family History Center had also made comments about my research success. As I began turning the handle of the Microfilm Reader, I started planning my strategy, and thinking about what to do next. I was ready to start rewinding the film when I noticed there was more information. I was overwhelmed with joy upon turning the Reader. There, on the last page of the 1820 census were the names of 16 Free Colored Persons, who were heads of households. There was a Thomas Day, a John Day, and a George Day included. Also listed, were the wives and children of the 16 households. There were 80 free people of color listed in the Census for Person County, North Carolina in 1820. Out of the 80, there were 16 with the last name of Day.

I had mixed feelings for the next several moments. It was a very emotional experience for me. I was very happy to find the three Day family households. It was, however, very sad to think about the 3,694 slaves who were listed in this census with no names, just numbers in the male and female slave columns. I was unable to do any more research that day. I kept thinking about those unnamed slaves.

1820 was the first year that Scott appeared on the U.S. Census. This was the same year that Thomas and George were first listed. Names for children, and women, were not given for that year of the census. I determined that several of the boys listed for Thomas and George Day fit the age category of Scott. However, I was unsure which one was Scott. He would have been about 15 years old in 1820, but children's names were not given in that census report. There was also a John Day, who was alone on the report, and had no children. There were check marks in the 1-14 age columns for Thomas and George, indicating that they each had sons in that age category. Now I needed to find out if Thomas or George was my great-great-great-grandfather. The three previous Census Schedules gave me no clues.

Tracing Scott's life was possible by following, and exploring the U. S. Census reports. Scott was born about 1803, in Person County, North Carolina. This was the year that the United States more than doubled its size through the Louisiana Purchase. Scott was born the same year as Philosopher, Ralph Waldo Emerson. This was also the same year that Ohio, (my birthplace) became a State.

On January 5, 1804, the Ohio Legislature took the lead in passing "Black Laws" designed to restrict the rights and freedom of movement of free Negroes in the North. I found records of a William Howard Day, who was very involved in doing away with these Black Laws. I have not connected him to my family, but I have enjoyed learning about his extraordinary life.

With the discovery of William Howard Day, I stopped my research for a while, to learn more about his interesting life. I have really enjoyed reading about his many accomplishments. I discovered the following books and articles about his life and times, **(See Appendix II).**

William Howard Day

Returning to my research, I learned that by 1816, when Scott was a young boy about 13 years old, the American Colonization Society was organized in Washington, D.C. It was formed to ease American race problems by transporting free Blacks to Africa.

Dora and I continued our search for Scott's father. While we wanted to find Scott's father, we were also interested in learning more about Scott, and his daughter Milly. We especially wanted to find out what they were doing, and what life was like for them during the Civil War period. The information found for Scott during this period has helped tremendously.

We began to learn more about what life was like in North Carolina during this period. From our early history lessons in school, Dora and I had both learned about the Civil War era. Now we were searching through the history of North Carolina, for information about our ancestors.

Our search for the father of Scott continued. It was a surprise to locate Scott on seven different census schedules. He was the connecting link that we needed to trace the family through so many decades. It appears that Scott lived past the age of eighty.

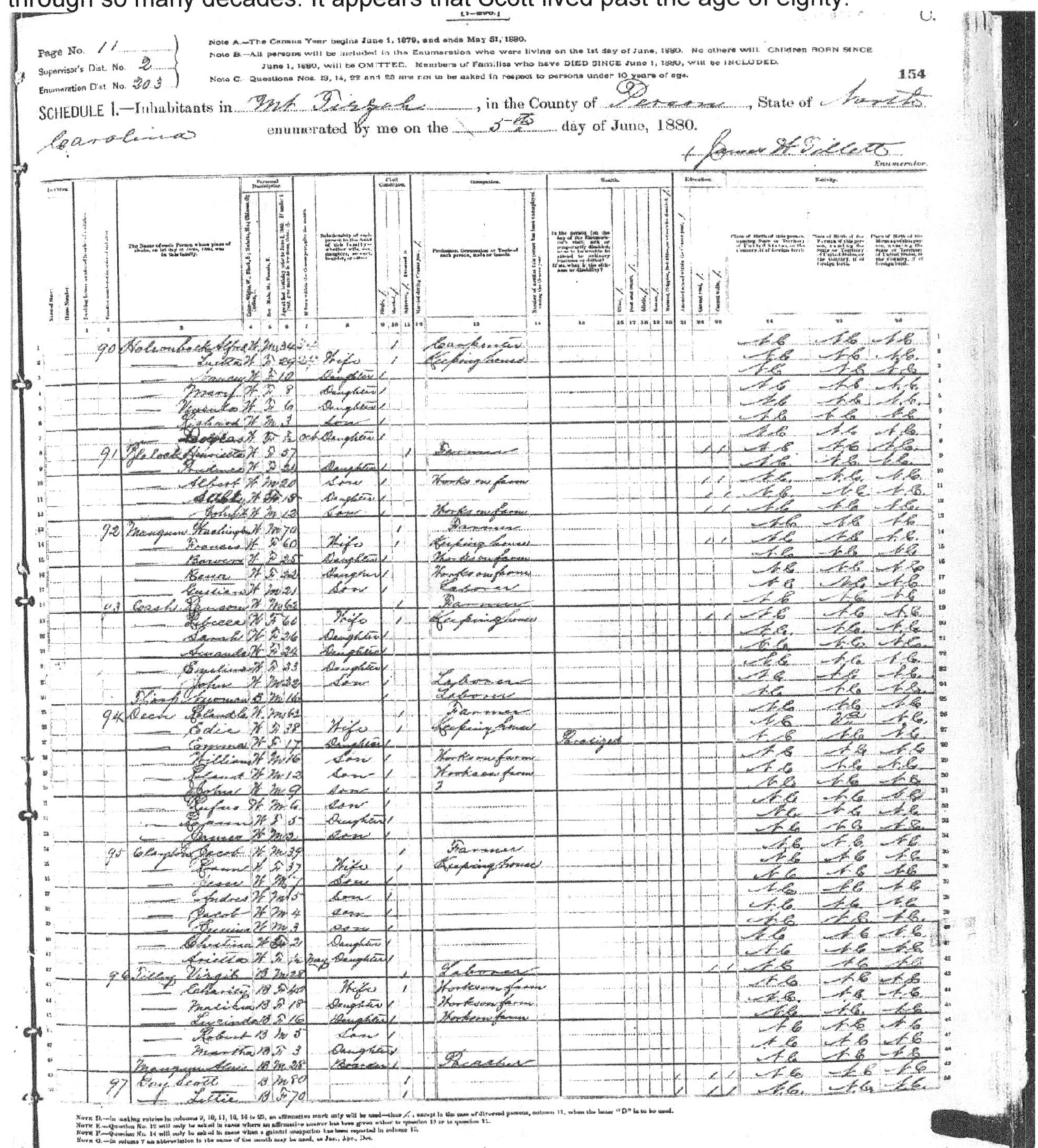

1880 CENSUS

**Scott Day was located at the very bottom of the 1880 Census with wife Setta (Lettie)
He was about 80 years old**

Page No. 13 ... Inquiries numbered 7, 16, and 17 are not to be asked in respect to infants. Inquiries numbered 11, 12, 15, 16, 17, 19, and 20 are to be answered (if at all) merely by an affirmative mark, as /.

SCHEDULE 1.—Inhabitants in _Flat River Township_, in the County of _Person_, State of _N. Carolina_, enumerated by me on the _12th_ day of _July_, 1870.

Post Office: _Roxboro N. Carolina_   James Wright, Ass't Marshal.

590

1870 CENSUS

Scott Day was located on the 1870 Census, with wife Setta & his daughter Milly (my g-grandmother), and her children (Aaron, Henry, & Setta)

Page No. 147

SCHEDULE 1.—Free Inhabitants in ______ in the County of _____ State of N Carolina, enumerated by me, on the 14th day of Sept 1860. W. H. Long, Ass't Marshal.

Post Office Hustonstone 517

Dwelling-houses	Families	The name of every person whose usual place of abode on the first day of June, 1860, was in this family.	Age	Sex	Color	Profession, Occupation, or Trade of each person, male and female, over 15 years of age.	Value of Real Estate	Value of Personal Estate	Place of Birth, Naming the State, Territory, or Country.	Married within the year	Attended School within the year	Persons over 20 yrs who cannot read & write	Whether deaf and dumb, blind, insane, idiotic, pauper, or convict.
1	2	3	4	5	6	7	8	9	10	11	12	13	14
1083	1081	Robert Torian	56	m		Farmer		10000	NC				
		Adelaide Torian	29	f					NC				
		Mary	12	f					NC				
		Robert	8	m					NC				
		Martha	5	f					NC				
		William	2	m					NC				
1084	1082	Josiah Bumpass	55	m		Farmer	1500	1500	NC				
		Elizabeth Bumpass	49	f					NC				
		Noah	23	m					NC				
		Sarah J	21	f					NC				
		Mary C	18	f					NC				
		Gideon M	11	m					NC				
		Hester	8	f					NC				
		Scott Day	36	m	m	Servant			NC				
1085	1083	Thomas Baxter	54	m		Farmer	2000	6000	NC				
		Lucretia Baxter	47	f					NC				
1086	1084	Nathaniel Torian	67	m		Farmer	14000	54000	NC				
		Martha Torian	60	f					NC				
		George L Torian	31	m		Farmhand			NC				
		Martha	25	f					NC				
		Mary Moore	45	f					NC			1	
		Robert Moore	12	m					NC			1	
		Nathaniel Dixon	10	m					NC				
1087	1085	S. M. Cunningham	68	f			9260	90760	NC			1	
1088	1086	A. K. Cunningham	37	m		Farmer	15200	154800	NC			1	
		M. K. Cunningham	22	f					NC				
1089	1087	W. H. Long	57	m		Farmer	6000	12000	NC				
		Jane Long	47	f					NC				
		A. H. Long	24	m		Mechanic			NC				
		Sarah D	20	f					NC			1	
		Reuben	19	m		Farmhand			NC			1	
		Joseph P	18	m		do			NC			1	
		George H	16	m		do			NC			1	
		Sarah J	13	m					NC			1	
		Elizabeth J	12	f					NC			1	
		John P	10	m					NC			1	
		Daniel	8	m					NC				
		David	4	m					NC				
		Elijah	2	m					NC				
1090	1088	Dr. W. Jordan	54	m		Physician	1200	26700	NC				

No. white males, 21   No. colored males, ___   No. foreign born, ___   No. blind, ___   No. idiotic, ___   No. convicts, ___
No. white females, 18   No. colored females, ___   No. deaf and dumb, ___   No. insane, ___   No. paupers, ___

1860 CENSUS

Scott Day was located on the 1860 Census
(14 from the top)

1850 SCOTT DAY

SCHEDULE I.—Free Inhabitants in ______ in the County of ______ State of ______ N Carolina enumerated by me, on the ___ day of Sept 1850. ______ Ass't Marshal

469

	Name	Age	Sex	Color	Profession, Occupation, or Trade	Value of Real Estate	Place of Birth				
793 755	Asa Vaughan	61	M		Farmer	900	N C				
	Mary	54	F				N C		1		
	Monroe	17	M				N C		1		
	Isabel	15	F				N C		1		
	Mary	13	F				N C		1		
	Asa	10	M				N C				
794 756	Cullin Bup	60	M	M	Husband		N C				
	Cyntha	40	F	M			N C				
	William	18	M	M	Husband		N C				
795 757	Mary Williams	70	F			187	N C		1		
	Susanah	55	F				N C		1		
	Sarah	45	F				N C		1		
	Sett May	45	M	B	Farmer		N C				
	Sarah Wyatt	12	F	M			N C				
	Charity Daniel	22	M		Cropper		N C		1		
796 758	Lydia Fowler	57	F				N C				
	Wilky	20	F				N C				
	Mildred	17	F				N C				
797 759	John Oakley	83	M		Farmer	250	N C				
	Elizabeth	50	F				N C				
	Tracia	40	F				N C		1		
	George	38	M		Illustrious		N C				
	Rebecca	36	F				N C		1		
	John	30	M		Cropper		N C				
798 800	Elijah Hopgood	54	M		Farmer	114	N C				
	Rebecca	36	F				N C		1		
	Ann	32	F				N C		1		
	Mary	25	F				N C				
	Eliza	22	F				N C		1		
	Louisa	19	F				N C				
	Charina	13	F				N C		1		
	Josephine	18	F				N C				
	Henry	6	M				N C				
	Washington	4	M				N C				
799 801	Allen Hopgood	38	M		Farmer		N C				
	Partheny	28	F				N C		1		
	Isaiah	9	M				N C				
	Sampson	5	M				N C				
	Virginia	1	F				N C				
800 802	Stephen Oakley	45	M		Farmer		N C				
	Nancy	46	F				N C		1		
	Diller	17	F				N C				

1850 CENSUS

Scott Day was located on the 1850 Census - (13 from the top)

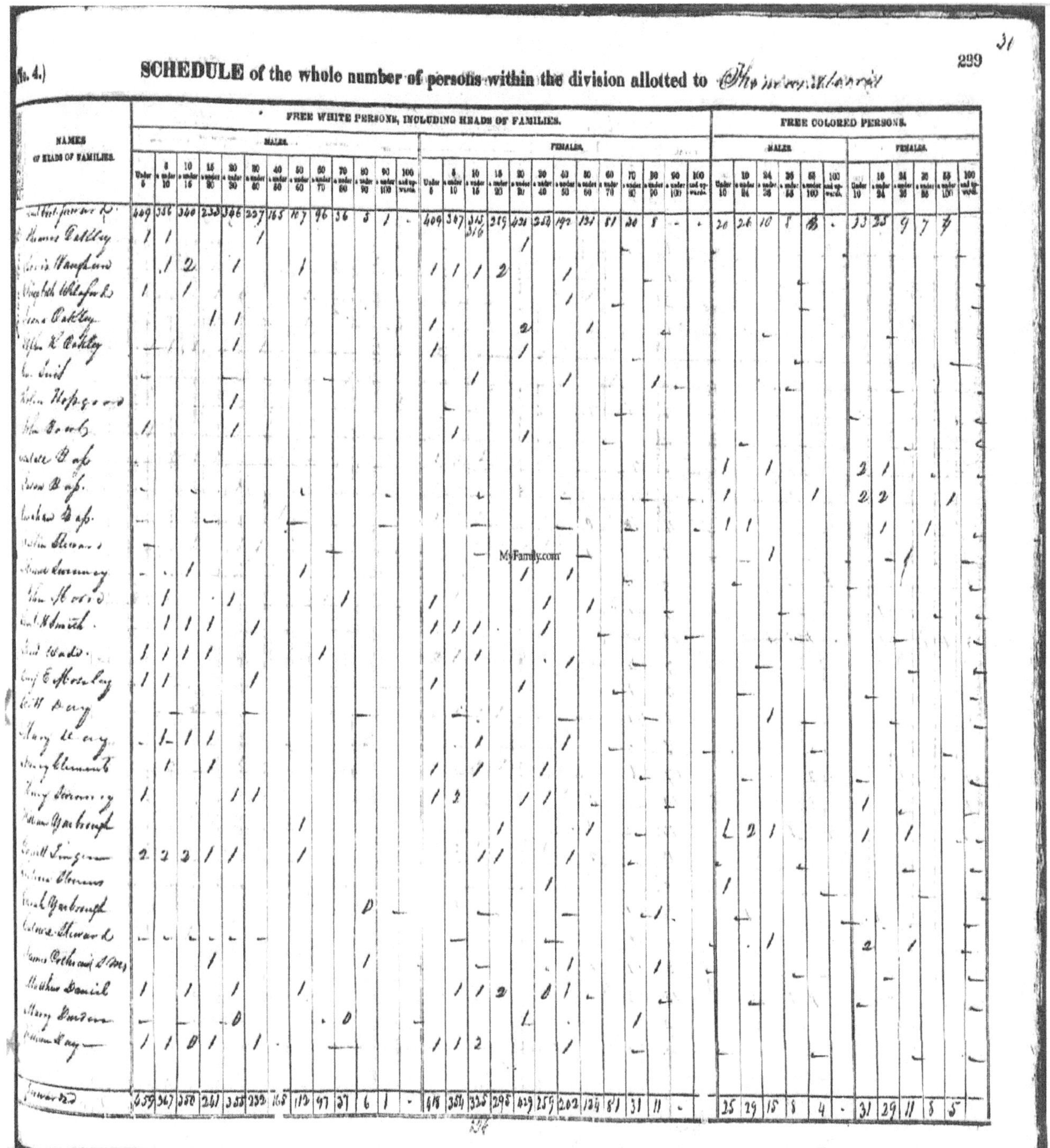

1840 CENSUS

Scott Day was located on the 1840 Census - (13 from the bottom)

CHAPTER 6

THOMAS R. DAY - GENERATION 6

&

RACHEL DAY – GENERATION 5

Thomas R. Day - Generation 6, My 3-G-G-G-Grandfather

After discovering Thomas, George, and John on the 1820 U. S. Census Schedule, I began checking earlier records to see if I could locate additional information about them and the Day family. This is how I made the connection to my great-great-grandfather Scott.

From "Caswell County Apprentice Bonds" by the noted historian, Katherine Kerr Kendall, I found Court records involving Thomas and George Day. The records indicate that Thomas, son of Rachel Day, and George, son of Ann Day were bound out as indentured servants to a Samuel Winstead on June 20, 1780. Thomas was three years old at the time, and George was four years old. I had now discovered the mothers of Thomas and George - another generation.

George was bound to Samuel Winstead on June 20, 1780. His mother Ann also had another child, a daughter named Lucy who was listed on these records. Lucy was bound to Drury Allen on March 20, 1780. Rachel also had another son. He was named Jesse (Jessie), and also bound to Drury Allen.

Fiber boxes CR020.101 and .07 at the Department of Archives and History, Raleigh, North Carolina, 28611 contain Indenture or Apprentice Bonds of Caswell County. Many of these Bonds are recorded in record books of the county. Caswell County was formed in 1777 from a part of Orange County. The county that Thomas and George lived in Person County was formed in 1791 from a part of Orange County.

 Finding this list by Kendall, was like discovering a gold mine. I uncovered so much information, some of which was very crucial to my research. The amount of detail that this one page contained was unbelievable. A review of this page shows that I discovered the following information;

 * **Rachel and Ann Day** - From this page of court records, I found two Day women. The possibility of a relationship between the two women would mean the discovery of additional family ancestors. I believed that either Rachel or Ann was my great-great-great-great-grandmother.

 * **Jessie Day** - From this page of court records, I found Jessie Day, the son of Rachel, and the brother of Thomas.

 * **Lucy Day** - From this page of court records, I found Lucy Day, the daughter of Ann, and the sister George.

 * **Thomas, George, Lucy, and Jessie** - From this page of court records, I discovered the ages of Thomas, George, Jessie, and Lucy.

 * **George Day -** From this page of court records, I found a George Day, who was bound out in 1820 to J. S. Hutchins, 12 July 1820. This could possibly be the son of the George Day, who was bound out in 1780 To Samuel Winstead.

* **Samuel Winstead** - From this page of court records, I found the name of Samuel Winstead. Thomas and George Day were both bound out to him as Indentured Servants on 20 Jun 1780. Finding his name meant that I might be able to discover more about the lives of Thomas and George.

*__Drury Allen__ - From this page of court records, I discovered the name of Drury Allen. Jessie (Jesse), the son of Rachel Day, and Lucy, the daughter of Ann (Anne) Day, were both bound out on 20 Mar. 1780 to Drury Allen. Finding his name meant I might be able to discover additional information about the lives of Jessie and Lucy.

* **From the "Person County Record Books',** I made another discovery. In December 1794, John Day, a free-born Negro boy age 5 years was bound to Charles Allen. He was to become an apprentice until he arrived at age 21. The age matches that of the John Day in the 1820 Census. There is also a Jesse Day, 7 years old, who was also apprenticed to Charles Allen on December 3, 1794.

These record books of Person County contain wills, inventories, and sales, settlements of estates, taxable lists, and letters of attorney. There are also apprentice bonds, bills of sales and other records. The books are in the State Archives, Raleigh, North Carolina.

I felt very fortunate that I was able to locate court documents, which contained such detailed information about my ancestors. The fact that they were free during this particular era was unexpected, and quite a surprise for me. Were my ancestors freed during or after the colonial period? From the book 'Finding A Place Called Home' by Dee Parmer Woodtor, I reviewed the different reasons they may have been free during that period. Some of the reasons include: They were Indentured Servants prior to the Institutionalization of Slavery, or they were the children of a free White Indentured Servant. They may have purchased their own freedom. It is possible they stole their own freedom by running away. They may have even passed or assimilated into the White world because they were light enough to do so. There are records of some people being released because they served during the Revolutionary War. Additionally, a slave owner may have granted them their freedom, through a process called manumission.

One of my goals was to find as much information about the Days of Virginia as possible, and to see if I could connect them with my Days in North Carolina. The Church of Latter Day Saints maintains one of the greatest collections of genealogical information in this country. It is known officially as the Church of Jesus Christ of Latter-Day Saints (or Mormons), and is managed by the LDS of Salt Lake City, Utah. Numerous books located at the center contained information that was especially helpful to me.

From Deeds of Caswell County, North Carolina in Book 'A' page 1, I found information about Samuel Winstead. Andrew Davidson of County of Orange, province of NC, to Samuel. Vinstead of Northumberland Co. VA, for 143 lbs., 205 A on Mayho adj. Thomas Mutter's line formerly Thomas Ring, John Beard formerly John Bridges. 20 Feb 1777. Test: Gabriel Davey, John Anderson, and James Davey (Also spelled Winstead).

It appears from this entry that Samuel Winstead was from Northumberland County, Virginia, and bought property in Caswell County, North Carolina in 1777. This was

three years before Thomas Day and George Day became his indentured servants in Caswell County, in 1780.

From Caswell County, North Carolina Land Entries 1778-1795 on page 1246, Dec. 21, 1779 - Aisley Winstead enters [omitted] AC on waters of Mayo Cr; border: Thomas Day and "others" - 'Woodland.'

Thomas Day and George Day, were bound out as indentured servants in 1780 to the father of Aisley, Samuel Winstead. It is unclear who this Thomas Day was, as Rachel's son Thomas was only two years old in 1779. Could he possibly have been the father of Rachel and Ann?

Through my research, I have learned that the free Negro apprentice was schooled in the art of making a living, and was usually taught to read and write. There was enacted in 1726, a law for binding out able-bodied Negroes who did not work. This law was to prevent vagrancy as a result of the increase in the numbers of free Negroes. As early as 1733, the children of free Negro parents were being bound out as apprentices. In that year, it was reported to the Upper House of the General Assembly that; certain free people, Negroes and Mulattos were bound out until they were 31 years of age. The person being bound out had no say in the matter.

There was a Virginia law that said: Mulatto children of White females were to be bound out as Indentured Servants for a period of thirty-one years, after which they were free. The law was later changed to eight years. There was also a law that bound out illegitimate children, for a certain number of years. The laws did not overlook White children, as they were also bound out as indentured servants, especially upon arriving in the new colony.

A single free Negro child was bound to an individual for the purpose of learning a specific trade. That Trade master in return, would provide essentially everything necessary to raise the child. The child in return, would work, and learn until a specified age. A law which passed by the Assembly in 1762, states that all free base-born children were to be bound out until they reached twenty-one years of age. The number of free Negroes in Orange County increased from 108 in 1800, to a maximum of 631 in 1840.

Thomas Day and George Day must have had a very difficult time during this period. Free persons of color were often despised, and viewed with contempt by a majority of the Whites, and looked down upon by many of the Negro slaves. Raising families must have been extremely difficult for Thomas, George, and their wives.

The following is information I discovered about Thomas Day, (who lived past the age of eighty) and his descendants in North Carolina. Thomas Day was born in 1777. This was the year that Vermont abolished slavery (July 2, 1777). Thomas spent his early childhood working for Samuel Winstead, and learning a trade. It is believed that Winstead, was a Blacksmith.

On April 12, 1787 when Thomas was about ten years old, Richard Allen and Absalom Jones organized the Free African Society. This was a Negro self-help group in Philadelphia.

It was decided in 1791, to form another county from Caswell, and that county was named Person County. This is where Thomas grew up. As was the custom after the Revolutionary War, as new counties were formed, many of them were named for men who had been prominent leaders during that period. When Caswell County was divided in 1791, Thomas Person's name was given to the Eastern half, Person County. The Day families lived near Thomas Person, before and after the War.

Thomas Day would have been a young boy 14 years of age in 1791. He and his family were probably not affected by the division of the county that year, as it was just a name change for their area. It is unknown how often he was able to see his mother Rachel and his brother Jessie during this period. They may have been living nearby. Thomas would still have been working as an indentured servant for Samuel Winstead.

Thomas and his family were probably very active within the church. Records reveal that Day ancestors were Founding Members of several Churches around 1890. These include: the Cedar Grove Baptist Church of Person County, North Carolina, and the Cleggs Chapel Baptist Church of Timberlake, North Carolina.

When Thomas was approximately 26 years old, Blacks in North Carolina presented a petition to Congress on January 20, 1797. They were protesting a State law that required slaves, although freed by their Quaker masters, to be returned to the state and to the status of slavery. The Congress rejected this first recorded anti-slavery petition by Negroes.

Person County, North Carolina has come a long way since her beginnings. The state has progressed tremendously since those early days. It was within this setting, that Thomas Day and his descendants carved a way of life. Thomas and his wife Mille were one of the pioneer Day families of North Carolina. The first time that Thomas appears on the U.S. Census was 1820 he was forty-three years old. A lot was going on that year. That was the same year that the Emancipator, the first Anti-Slavery magazine was published. It was issued monthly from April 30, to October 31, 1820. Elihu Embree was the Editor and Publisher of the magazine. Also, on March 3, 1820, Congress approved the famous Missouri Compromise. Slavery was prohibited North and West of the 36-30 parallel line within the Louisiana territory.

Earlier records do not always indicate whether a person was Black or White. However, I was able to find information about Thomas Day in 1824 Court Records for Person County, North Carolina. Thomas would have been approximately 47 years old at that time.

From Deed Book 'G' for Person County, North Carolina, the following entry was found: Thomas Day, a coloured man, (in debt to Colonel John Day for $14) to Isaac Day, trustee, a gray mare with one eye about 13 years old, 6 Jan 1824. Wit: Philip Obriant.

Deed books for Person County are at the office of the Registrar of Deeds, Roxboro, NC, 27573. Books 'A' through 'G' total 2,899 pages covering 33 years of land transfers, some grants from the State of North Carolina, deeds of slaves and other personal property, powers of attorney, division of land to inheritors, and migration of inhabitants. Other records have been found for John and Isaac Day, who both appeared on the 1800 U.S.

Census Schedule for Person County, North Carolina. I have been unable to determine if they were related to my ancestors.

Continuing my search, I began to uncover more information about the life of George Day, who was the son of Ann Day. Ann also had to have her son George, bound out as an indentured servant at age four. As noted, he was bound out to the same person that Thomas Day was bound to, Samuel Winstead. According to county record books, George Day was born in 1776. The 1820 Census shows that he was married however wives and children's names were not given on that census. George was born during the period of the Revolutionary War. This was the time the early colonies were beginning to fight for independence from England.

The American Revolution was a period of progress and reform. When the revolution began, the Church of England was the official church in many states. Everybody, no matter what he believed, was taxed to support the official church. After the revolution, church and state were separated, and people were allowed to worship as they pleased. The new United States came out of it very well, and General George Washington was elected our first President.

George Day would have been about 13 years old in 1789, still living on the Samuel Winstead estate, and working as an indentured servant. The years 1776-1789 were considered the 'Experimental Period' in American History. This was the time when Americans were experimenting and discovering their nationality. America had finally won independence from England, and was trying to establish itself as a Nation.

At the end of the 'Experimental Period' in 1789, Benjamin Banneker accurately predicted a Solar Eclipse. Like George Day, Benjamin's grandmother had also been an indentured servant. After her indenture period, she bought property, and eventually was able to provide a good life for her family. Benjamin had the privilege of growing up free, and receiving a great education. He was an outspoken critic of slavery, and the Revolutionary War.

It is difficult to imagine how George was affected by the Revolutionary War. Like most people of the time, he was probably glad the fighting was over. The war had been going on for all of his young life, so that peaceful times were completely unknown to him. In later years, George and his wife had a total of at least ten children, perhaps more. The 1820 Census lists a wife, three sons, and three daughters. The 1830 Census shows two sons, and two daughters, all less than 10 years of age. There is also a daughter 24-26 years of age. The 1830 Census is the last year that George appears his age is listed as 55-100. He would have been 54 or 55 years old, because he was four years old when he was bound out in 1780, one year older than Thomas.

I continued searching through records to learn more about my ancestors. From Caswell County Court Records, March 1783, the children of Ann Day and Rachel Day were now orphans. Lucy, the daughter of Ann, who was bound out in 1780 to Drury Allen, is now being bound to his son David Allen. Lucy is now four years old. The same court record shows that the children of Rachel are now orphans, and they are bound to David Allen. Jessie, who was also bound out to Drury Allen in 1780 at six months, is now bound to David Allen.

From Person County, North Carolina Deed Books – Deed Book A, State of NC - #1020, Aisley Winstead for 50 Shares per 100 A. for Acres adj. Gabriel Davey, Thomas Day, 18 May 1789.

Aisley Winstead, and the older Thomas Day are still neighbors ten years later. The county of Person was formed in 1791, out of a portion of Caswell County. This 1789 record, although listed in the Person County Deed Books, was probably a record of Caswell County, North Carolina.

From Person County, North Carolina Tax List by Districts for 1794, the following records were located. The Nash District shows Charles Allen with 804 acres, 1 White poll and 2 Black polls. David Allen had 1,000 acres with 1 White poll, and 2 Black polls. Drury Allen, Sr. was listed with 900 acres, with a zero in the White poll column, and 2 Black polls. Samuel Winstead was also in the Tax List for Nash District in 1794. He had 450 acres, zero in the White polls column (possibly because of his advanced age), and 3 in the Black polls column. Two of the Tax polls were probably for Thomas Day, who would have been seventeen, and George Day, who would have been eighteen years old in 1794.

I was curious about the older Thomas Day who was living in the neighborhood, when Rachel and Ann had their sons bound out to Samuel Winstead as indentured servants in 1780. It is doubtful that he was the father of either Thomas or George, because he was still living near Aisley Winstead in 1789. The children of Rachel and Ann were listed as orphans in the 1783 court record, and this Thomas was still living.

Looking through court records, land and property deeds, and other documents, I began to locate more information about this Thomas Day. I also discovered an enormous amount of information about Day families living in the area at the time. In many of these records, race is not listed, so it is difficult to determine which ones are my ancestors. I began to list these records by date order in an effort to learn more about the Days, and their neighbors (See Appendix I).

I felt good that I was learning so much about my ancestors. When I started this project, I never imagined that so many records would be available about our past history. I began wondering if it would be possible to go back to the beginning of the century.

In November of 2008, I received an e-mail from a Natalie Pryor that helped me tremendously with my Day family research. The information she sent to me solved several puzzles regarding my family history research.

Hello Mr. Day,

I discovered your website (www.day-banks.com) this week and was grateful for the information provided on searching for free Blacks prior to the Civil War. In reading about the information you listed about your own family search, I also discovered that it is more than possible that we are related.

My name is Natalie Pryor. My mother, Camilla Pryor is the daughter of Martha Torain-Webb and Ollie Webb. Ollie Webb is the son of Charlotte Day-Webb and Robert L. Webb, and Charlotte's parents are Eliza and Richmond Day. On the 1880 Census, it lists

Richmond Day at age 32, living with wife, Eliza, daughter Charlotte, sister in-law Catherine and a cousin James at age 14 or 19. Could this be the James L. Day in your family line?

If so, I am very curious about the records you found concerning George Day. George and Emily Day are the parents of my great-great-grandfather, Richmond Day. The information you found about George, his mother Ann and the fact that George and Thomas were bound to Samuel Winstead in 1780 as apprentices is extremely interesting to me.

I live in New York, and am not able to travel to North Carolina easily. I am writing you in the hopes that you might provide information that may not be included on your website, and might shed some light on my family's background.

Thank you and warm regards, Natalie.

Natalie and I are definitely cousins. Thanks to Natalie, I now have information that reveals that Rachel Day was my 4-great-grandmother, and her sister Ann, was Natalie's 5-great-grandmother. My grandfather, James L. Day, was 19 years old on that census. Natalie and I have been e-mailing each other and keeping in touch. I am very happy that Natalie got in touch with me, and I have been able to learn more about that side of the family. From my research, it seems like the first cousins of our family through each generation, have always kept close contact with each other. That is also especially true of my maternal ancestors, as well as my paternal ancestors.

My cousin Natalie sent me the following additional information regarding her research.

DAY HISTORICAL NOTES - from Natalie Pryor

John Day, Sr., was born in 1766. By 1820, his family and other branches of the Day family migrated into North Carolina. Many free Black Days settled in Orange County, adjacent to Caswell County, and were well established there in the 1830's and 1840's. Caswell County is located in the Piedmont region of North Carolina next to the Virginia border, and began as the Northern region of Orange County, North Carolina. Established during the American Revolution, Caswell County was the first county to be ratified by the North Carolina General Assembly in 1777. (Thanks for the information Natalie). My research for Rachel Day follows.

Rachel Day - Generation 5, My 4-G-G-G-G-Grandmother

This particular story reviews the Day family history back to Rachel Day. Rachel was my great-great-great-great grandmother. I researched back through the generations to locate Rachel and to tell her many descendants about the Day family history. Rachel was born approximately 1760, and her descendants up to the year 2009, total nine generations. Rachel would probably be as proud of her descendants, as they are of her. My great-grandsons A'Sean D'Anthony, and Tyray are members of the tenth generation.

The era that Rachel lived in was during the Revolutionary Wartime. This was the time the early colonies were fighting for independence from England. When the American Revolution began, the Church of England was the official church in most areas. During that time period, everyone was taxed to support the church. However, after the Revolution, people were able to worship wherever they wanted, and the church and state were separated. Like most of the population, the Revolution must have affected Rachel and her family tremendously.

Rachel lived in Caswell County, North Carolina in 1780. Caswell borders the Southern part of the State of Virginia, and was originally a part of Orange County, North Carolina. Caswell is also in the Piedmont plateau region, west of the coastal plain. Quite a large number of Rachel's many descendants are still living in the Piedmont region of North Carolina.

During that era, it was not uncommon for children to be indentured without the consent of their parents. However, children were mainly indentured out to learn a trade, so that they could eventually earn a living for themselves. The trade master in return would provide everything necessary to raise the child. It was discovered that several of Rachel's children were indentured out. Rachel was probably very aware of the advantages for her children to be able to learn a trade, as survival must have been a high priority. This system of training was highly popular during that period. However, many times the period of indenture for some of the free people of color, was very long.

There were also other forms of indenture. It is believed that the first African Americans to come to Virginia were indentured servants. Also, in the colonial days, people who did not have the money to come to this country would often bind themselves out as indentured servants. They would work for a specified number of years for the person who paid their way, and then they would be free to work for themselves.

As explained earlier, in a 1780 court record, Rachel and her sister Ann were located with their children. Rachel's son Thomas was being indentured to a Samuel Winstead and her son Jesse was being indentured to a Drury Allen. Ann's son George was also indentured to Samuel Winstead, and her daughter Lucy was indentured to Drury Allen. It was apparent, that the Winstead and the Allen families needed further investigation.

The lives of the Day families, the Allen families, and the Winstead families seemed to be mysteriously intertwined during that time period. The three families, as well as the neighboring families have been followed and researched in an effort to learn more about Rachel and her children. The will of a Samuel Winstead, indicates the family was living in Northumberland, Virginia in 1720, and were members of the St. Stephens Parish.

In the 1720 will, Samuel Winstead was leaving property to his son Samuel Winstead, and to his grandson, also named Samuel Winstead. A land entry record for Samuel Winstead, the grandson, reveals that he bought property in Caswell County, North Carolina in 1777, and was from Northumberland County, Virginia. This was 3 years before Rachel's son Thomas became his indentured servant.   Court records show that the Day family was also associated with St. Stephens Parish as early as 1692.

This was a very difficult time for Blacks, those who were free and those who were living in slavery. Rachel and her family must have paid some very heavy dues during that period, just to survive. It was especially difficult to keep a family together during those years. Like most of the free Black families of that time, Rachel and her family were probably constantly on guard to preserve what freedom they had. Records seem to indicate that a majority of the free Black families in North Carolina lived in the rural areas.

Many free Blacks were sheltered and protected by White families, and were therefore able to maintain their freedom. In 1785, an Act by the Legislature required the use of badges for Negroes. Free Negroes had to register with the Town Commissioners, and pay a fee for their badges, which indicated that they were free. If caught without their badges, they could have been picked up and sold into slavery. Free Negroes were often stolen and sold into slavery. There was a law passed in 1779 against the stealing of free Negroes.

There were many pioneering Black Day families living in the North Carolina area long before 1779. In an early colonial record, a passage was found referring to a Henry Day, who lived in Granville, County, which borders Caswell to the East. Granville was also originally a part of Orange County. It seems that Henry was one of the hunters that the colonialists used to assist them in finding the different bodies and tracts of land. Henry was one of the men who worked with the surveyors who measured the boundaries and tracts of land in North Carolina.

The first U. S. Census in 1790 shows that there were four families named Day, living in Caswell County in the Hillsborough District. They were John, Isaac, Henry, and Francis Day. Isaac Day was the only one not listed twenty years later in the 1810 census. There were seven-Day families by then, which included John, Francis and Henry's sons. The three sons were listed as juniors, and the fathers as seniors. There was also a third Henry Day listed, without junior or senior behind his name

From Land and Deed entry records for Person County, North Carolina, details are given of the Day families living there. Person County was formed from a portion of Caswell County in 1791. From colonial records, a petition to Governor Josiah Martin to form the county of Person was signed by a John Day, and a Thomas Day. Person County was named for Thomas Person, who was in the Revolutionary War. Day families were found to be living near him before and after the War.

It seems the Days were instrumental in helping to clear the land in North Carolina during the early colonial period. Areas had to be cleared in order to build homes, and to make roadways that were passable for wagons. Records indicate that the Days were part of the crews that carried and operated the equipment used to cut down trees and clear the forest areas.

There have been several land entry records located for a Thomas Day who lived in Caswell County, probably the one who signed the petition to form the County of Person. A 1779 record shows this older Thomas Day living near Aisley Winstead, the son of Samuel Winstead. Rachel's son Thomas was only 2 years old in 1779. As indicated earlier, a later 1789 record indicates that Aisley Winstead and Thomas Day were still neighbors ten years later.

It appears that the Winstead and the Allen families may have indeed been helping Rachel and her children to survive. In January of 1783, Rachel gave birth to a daughter named Nancy. This was just two months before Rachel died. In court records of 1783 Rachel and

Ann's children were listed as orphans and being indentured out. Rachel and Ann must have died several months before the court record, because Nancy was only two months old. The record shows that Rachel's son Jesse and daughter Nancy were being indentured to David Allen, the son of Drury Allen. Ann's Daughter Lucy was also being indentured to David Allen. This record seems to be another indication of the close ties between the Allen and the Day families. Evidently, Thomas and George were still indentured to Samuel Winstead.

The indenture of her children to David Allen after her death is probably what Rachel would have wanted. Like most mothers, she would have been concerned for the safety and welfare of her children. Later records indicate that Rachel and Ann's children did well, and continued to add to the Day family heritage. The Day family tree continues to grow with the descendants of Rachel - now nine generations. Rachel Day would certainly be very proud of her descendants. They are all proud to be on – the Day family tree, and will make sure that their descendants know the story of, Rachel Day, and the Day family history.

Boys sawing tree
Photo Credit: Abby Aldrich Rockefeller Folk Art Center, Williamsburg, Virginia

Many records have been found of the Day ancestors in the early colonial period. These records indicate that many free people of color, including the Day ancestors worked to clear the areas so that roads could be developed. Various documents found in Person County, North Carolina listed free Day ancestors working for Thomas Person and his family. Person County, North Carolina was named for Thomas Person, of Revolutionary War fame.

NASH DISTRICT, PERSON COUNTY (Continued)   *1794 TAX LIST*

Name	Land in Acres	White Polls	Black Polls	Name	Land in Acres	White Polls	Black Polls
Winsted, William	100	1	0	Winsted, Samuel	450	0	3
Wheeler, Samuel	100	1	2	(1 stud horse)			
Wilson, John	204	0	0	Williamson, Henry (in			
Walker, Jesay	0	1	1	St. Luke's District)	194	1	1
Winsted, Constance	250	1	0				
Wilson, James, Esq.	200	1	2	Yarbrough, Samuel	363½	1	6
Winsted, John	0	1	0	Yarbrough, John	832	1	0
Wilkerson, John	150	1	0	(1 stud horse)			

End of Nash District - signed by Isaac Vanhook.

ST. JAMES DISTRICT

Name	Land in Acres	White Polls	Black Polls	Name	Land in Acres	White Polls	Black Polls
George Anderson	100	1	0	John Cash	904	1	0
James Anderson	100	1	0	Moses Cash	119	1	0
				Aaron Cash	0	1	0
John Bumpass, Sen.	500	1	2	Jno. Coughran	262	1	0
John Bumpass, Jun.	0	0	0	(1 stud horse)			
Jno. Bumpass	0	0	0	James Coughran	60	0	0
Milr. Blalack, Sen.	550	1	0				
Milr. Blalack, Jun.	0	1	0	Gab. Davey	204	1	3
Tho. Blalack	200	1	0	Isaac Day	173	1	0
Jno. Beardson	100	1	0	Frans. Day	144	1	0
Allin Burton	200	1	0	Henry Day	200	1	0
Dan. Burton	0	1	0	Mills Durdon	200	1	0
Sam. Bumpass	0	0	0	Elijah Denby	820	1	4
John Boles	140	1	4	James Daniel	312	1	6
Rich. Bland	945	0	6	Matthew Daniel	200	1	1
Wm. Blalack	550	1	0	Chs. Davis	0	1	0
				Robt. Dickins	13432	3	15
John Cliaby	840	1	0	(4 wheels)			
Dan. Coleman	571	2	0	(1 stud horse)			
Harey Crews	1031	1	0	Jefre Dickins	1600	1	2
Dan. Clayton, Jun.	0	1	0	(2 wheels)			
Jno. Clayton, Jun.	240	1	0	Willm. Duncan	200	1	0
John Cate	0	1	0	Robert Davis	0	0	0
Robt. Cate	0	0	0	Philip Day	0	0	0
Jno. Commins	0	0	0				
Petr. Cazorte	100	1	0	Dan. Farmer	178	2	0
Benj. Cazorte	119	1	0	Ste. Farmer	151	1	0
Jno. Cazorte	100	1	0	Nathanl. Farmer	150	1	0

7

1794 Tax List

**Cousins George and Thomas Day would be listed under the
Black Polls column for Samuel Winstead - (they would be in their teens)**

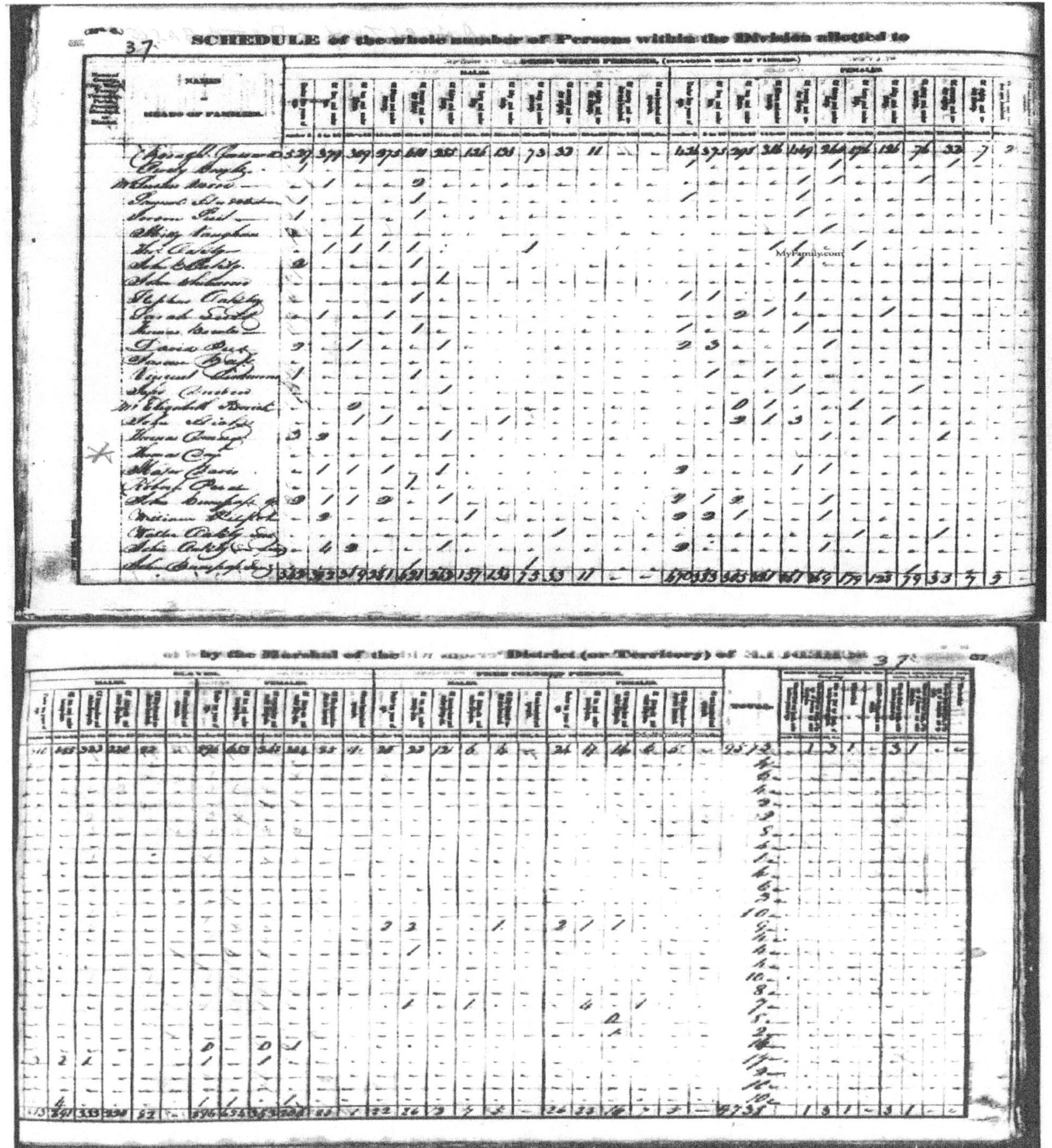

1830 CENSUS

Scott Day was located on the 1830 Census. The left half of the 1830 census is at the top. The name of Thomas Day is listed #8 from the bottom.

The right side of the census is at the bottom. All slaves and free people of color are listed on the right side of census. Thomas Day is the father of Scott, who is not listed by name, but should be about 25 yrs. old. Scott is listed in that age column.

1820 Person County, North Carolina Census

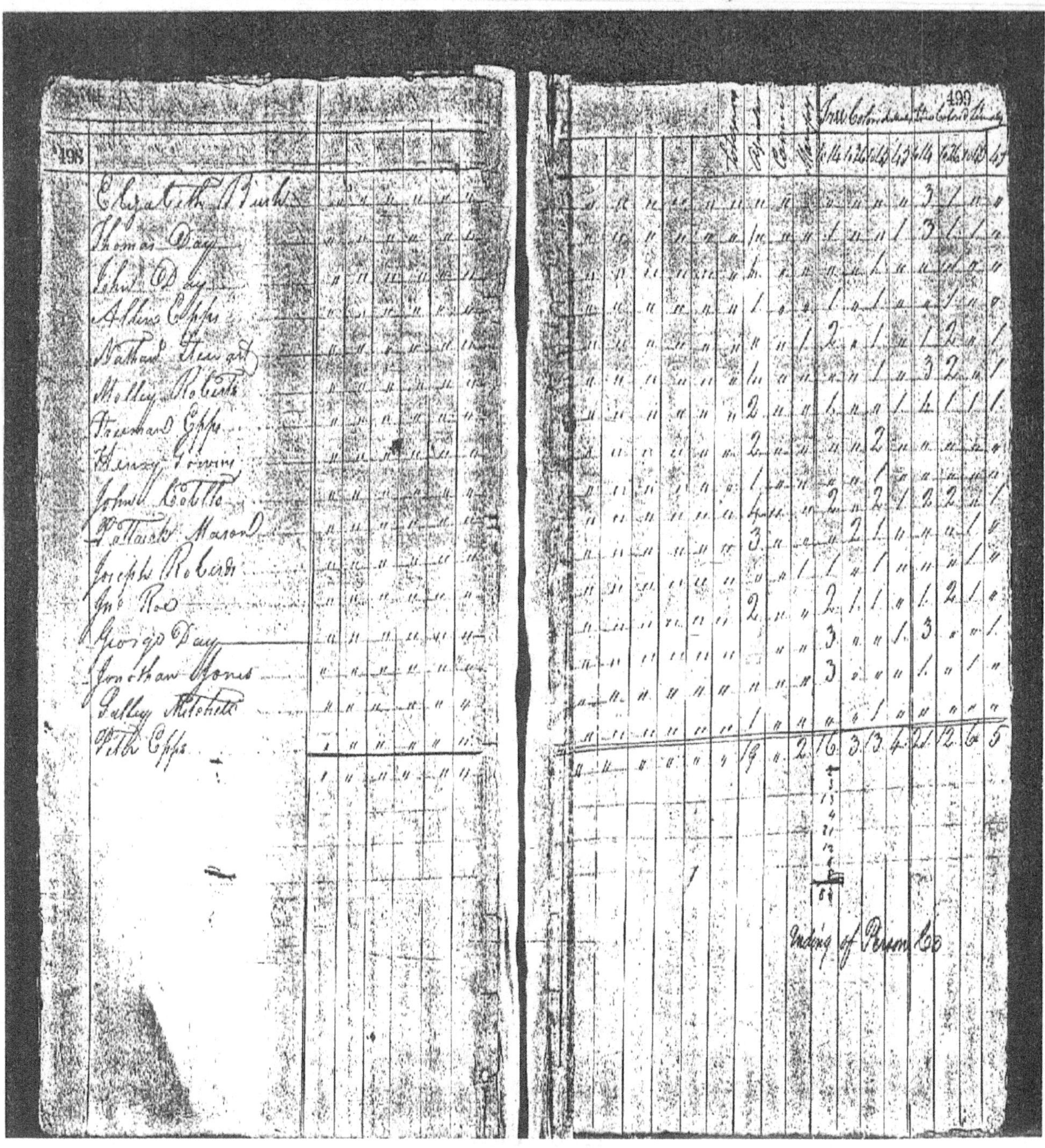

**Scott Day was located on the 1820 Census - Person County, NC
Scott is in the column for 15 year olds. He is with his father Thomas, and family.**

**Includes: Cousins John & George Day, & All 16 Free Colored Families In 1820
(This was the very last page of the Census Schedule)**

CHAPTER 7

THOMAS DAY, I. – GENERATION 4

WINNIFRED DAY (WEBB) – GENERATION 3

RACHELL DAY (WEBB) – GENERATION 2

MARY DAY & DANIEL D. WEBB – GENERATION 1

Thomas Day, I., - Generation 4, My 5-G-G-G-G-G-Grandfather

This chapter introduces the first four-generations of the Day family. They were traced back to the 1660's. What is known about them has been discovered through the following documents.

- **Church/Parish Records**
- **Wills**
- **Apprentice/Indenture Bonds**
- **Early Tax Records**
- **Land Entry Deeds**
- **Court Records**
- **State/County Records**
- **Early Birth Records**
- **Record/Order Books**
- **Estate Records**

With these documents, I have been able to connect the four-generations together, in the State of Virginia. It was the fourth generation (Thomas Day, I., and his family) that began the migration from Virginia to North Carolina, in the late 1770's. There were however, other Day families who had already located in North Carolina before Thomas and his family. It appears that the Day family traveled to North Carolina with the Samuel Winstead family. It was in 1777, that Samuel Winstead bought property in Caswell County, North Carolina, three years before Thomas R. Day was indentured to him, in a 1780 Court Record. Early records show that the Day and Winstead families were associated together at the St. Stephens Parish in Northumberland County, Virginia since the early 1690's.

In January of 1783, Rachel gave birth to a daughter named Nancy. This was just two months before Rachel died. In court records of 1783, Rachel and Ann's children were listed as orphans and being indentured out. Rachel and Ann must have died several months before the court record, because Nancy was only two months old. The record shows that Rachel's son Jesse and daughter Nancy R. Day were being indentured to David Allen, the son of Drury Allen. Ann's daughter Lucy was also being indentured to David Allen. Both Jessie and Lucy were indentured to Drury in the 1780 Court Record. This record seems to be another indication of the close ties between the Allen and the Day families. Thomas and George were still indentured to Samuel Winstead.

As mentioned earlier, several land entry records were located for a Thomas Day that lived in Caswell County. A 1779 record shows this Thomas Day, I., living near Aisley Winstead, the son of Samuel Winstead. Rachel's son Thomas was only two years old in 1779. A later 1789 record indicates that Aisley Winstead and this Thomas Day were still neighbors, ten years later, and this was after Rachel and Ann died in 1783. This Thomas was still living, and the children of Rachel and Ann were considered orphans. It appears that the Winstead, and Allen families may have indeed been helping Thomas, Rachel, Ann, and their children to survive. Records show that Thomas Day, I, was born in 1735, in Southampton, Virginia, to a Winnifred Day.

The year before Thomas was born, the Great Awakening, a Religious Revival starts in Northampton, Massachusetts. Prince Hall, Abolitionist Leader and the founder of African-American Freemasonry, was also born in 1735, the same year as Thomas.

It appears that some of his children were born in Southampton County, and others were born in Northampton County, Virginia. Researched documents show that Thomas may have possibly been father of at least eleven children.

- **Ann Day**
- **Winifred, T. Day**
- **Rachel Day**
- **John S. J. Day**
- **Jesse S. Day**
- **Joseph Day**
- **Benjamin Day**
- **Judith Day**
- **Nancy Day**
- **Stephen Day**
- **Samuel S. Day**

I was able to connect Thomas Day, I., to his mother Winnifred Day (Webb) through court records obtained from Southampton, Virginia, and Order Book 1699, 1713, Part 2, 809.

Winnifred Day (Webb) - Generation 3, My 6-G-G-G-G-G-G-Grandmother

St. Stephens Parish records revealed the following information: 17 Dec 1712 Rachell Day a Mulatto having a child named Winnifred. It is therefore hereby covenanted & agreed by the said Churchwardens that the said child named Winnifred born in June last for the consideration of the said Sanders his allowing the said Winnifred during the said term sufficient apparel, diet, washing & lodging saving the said Parish harmless shall serve & she is hereby bound to serve the said Sarah Webb her heirs & assigns from the day of the date hereof until she, the said Winnifred shall attained to the age of thirty-one years according to law in witness of premises the said parties have set their hands & signed sealed & delivered in the presence of ye Justice & acknowledged in Northumberland County Court by Capt. Richard Hues & Mr. Edwd. Coles, Churchwardens of St. Stephens Parish the 19th Feb. 1712/13 & is recorded. RB 1710-13, 270.

The following day, there was an additional notation. 19 Feb. 1712/13-Winnifred Day (alias) Webb her Indenture for service to Sarah Webb on ye motion of Edwd. Sanders is admitted to Record. OB 1699-1713. Part 2, 809.

In 1711, the year before Winnifred was born, Quakers persuaded the Colonial Government to outlaw slavery in Pennsylvania, but the law was vetoed by the British Government. In 1712, in New York City, a slave revolt took place, and nine whites were killed. There were over 20 black rebels killed during this revolt.

Records reveal that, at 23 years of age, Winnifred became the mother of Thomas Day, I, in Southampton, Virginia, 1735. At 28 years of age, she became the mother of John S. Day, born 1740. At 41 years of age, she became the mother of George Day, born 1753. This information was obtained from Order Book 1699-1713, Part 2, 809.

(1) Thomas Day, I., had the following family (see previous page, 4th generation).

(2) George Day had the following family.

   * **Jane Day - born Northumberland County, Virginia, in 1775.**

   * **Willoughby Day - born Northumberland County, Virginia, in 1778.**

(3) John S. Day had the following family

   * **Lucy J. Day - born Northampton County, Virginia, in 1760.**

   * **John J. Day - born Northampton County, Virginia, in 1764.**

From the noted historian, and author Paul Heinegg, I located information about John Day, Jr., and Thomas Day. Heinegg's books and research have been very helpful to me in my research. His reference books are excellent, and an invaluable resource for researchers. From the research he has done, I have been able to learn a great deal of information about my free Day families. His book entitled, **'Free African Americans of North Carolina and Virginia'** was a tremendous aid to me during my early years of research.

Like John Hope Franklin, Paul Heinegg's research through the years, and his books have been extremely helpful to me. I found many of my Day ancestors in his books, and contacted him about my research. We have been in contact with each other through numerous e-mails, and phone calls.

I learned that John J. Day married Mourning Stewart, daughter of Dr. Thomas Stewart of Dinwiddie County (Dinwiddie County Chancery Orders 1832-52, I; Sneed & Westfall, History of Thomas Day, 6). He was taxable in his own Dinwiddie County household in 1800 with his brothers-in-law, Henry and Armistead Stewart (Virginia Genealogist 18:185). In 1807, he purchased a small plantation in Sussex County where his children went to school (Sneed & Westfall, History of Thomas Day, 10). He was listed as a cabinetmaker in the Dinwiddie County list of 'Free Negroes and their occupations' between 1814 and 1817 (Personal Tax List). In 1820, he was head of a Warren County, North Carolina household of 5 'free colored' (NC:808). He died at the age of 68 years (Sneed & Westfall, History of Thomas Day, 10). His wife Mourning, was counted as an 84 year old woman in the 1850 Census for Caswell County, North Carolina. At 32 years of age, his son John J. Day, Jr. was born in Greenville County, Virginia, and on February 18, 1797. His son Thomas J. Day was born 1801.

The two sons of John J. Day, would one day become - world famous, and written about for generations to come. The Day brothers, John J. Day, Jr., and Thomas J. Day made such an impact on this country, and they contributed so very much.

*   **John J. Day, Jr.,** was born in Greenville County, Virginia, February 18, 1797. Like his father, John was also a cabinetmaker, but later became a Baptist minister. He emigrated to Liberia in 1830 with his wife and four children. He was Superintendent of the Baptist Mission, a signer of the Declaration of Independence, and became the second Chief Justice of the Liberia Supreme Court (African Repository 35 (1850):158; and 37 (1861): 154-58 by Wiley, Slaves No More, **(See Appendix III).**

*   **Thomas J. Day,** was born 1801. Thomas operated the Union (Yellow Tavern) in Milton, Caswell County, North Carolina. It was the third largest furniture factory in the State, from 1823 to 1858. He married Acquilla Wilson in Virginia in 1829, but needed special dispensation from the Legislature to allow her to migrate into North Carolina, since an 1827 North Carolina law made it illegal for the free African Americans to enter the state. He integrated the town's Presbyterian Church by replacing the worn pews on the main floor of the church in return for the right to sit in them, rather than in the gallery. The Yellow Tavern is a National Historic Landmark. His work is also in the North Carolina State Museum. He was listed as a 49 year old in the 1850 Census for Caswell County. His land near the main road from Milton to Yanceyville near County Line Creek was mentioned in an April 3, 1851 Caswell County Land Entry (Entry no. 1418). At the 2000 Day Family Reunion in Raleigh, North Carolina, I had an opportunity to visit the Museum, and also the Yellow Tavern.

I was able to connect Winnifred Day (Webb) to her mother Rachel Day (Webb) through court records, and Order Book 1699-1713 from St. Stephens Parish of Northumberland County, Virginia.

Rachel Day (Webb) - Generation 2, My 7-G-G-G-G-G-G-G-Grandmother

This was an extremely harsh period for free people of color who were living in America. A year before Rachell was born (1691), the existence of free blacks in Virginia was seen as a threat by the white colonists. Therefore a strict law to restrict manumissions was passed. In 1692, the State of Maryland passed a law that white men who marry, or have children by black women must spend seven years in servitude. There were also strict penalties for black men who had sexual relations with white women.

According to an Order Book from St. Stephens Parish of Northumberland County, North Carolina, Rachell and her twin brother John were born in 1692, and listed as Mulattoes (of mixed parentage). Rachell was located again in several court records about 17 years or so later. *On September 1709 - Rachell, a Mulatto girl belonging to the estate of John Webb deceased, being presented to this Court for having a bastard child by Arthur Thomas as she hath made oath in Court. It's therefore ordered she serve her master's estate according to law after her time by custom or law shall be expired. OB 1699-1713, Part 2, 619.*

Three years later Rachell is located in Court Records again. On *17 Dec. 1712-Rachell Day, a Mulatto servant to the orphans of John Webb deceased, having had a bastard child (Winnifred) by Negro Will (Capt. Kenner's Slave) as she hath made oath in Court. It's ordered she serve for her said default according to law. OB 1699-1713, Part 2, 803.* Regarding this same birth, there was another entry into Record Books several months later. *18 Feb. 1712/13 - This Indenture made the eighteenth of Feb. Anno Dom 1712, 13 between Richard Hues & Edward Coles Churchwardens of Saint Stephens Parrish in Northumberland County of the one and Edward Sanders one of the Executors of John Webb deceased for & in behalf of Sarah Webb, the daughter of the said John Webb deceased of the other part.*

Witnesseth that whereas by act of Assembly it is provided that if any woman servant shall have a bastard child by a Negro, the Churchwardens of the Parrish where such child shall be borne shall bind the said child to be a servant until it shall be thirty one years of age and Rachell Webb (alias) Day, servant to the said Sarah Webb having had a bastard child named Winnifred,   by a Negro pursuant to the said law.

I was able to connect Rachel Day (Webb) to her mother Mary Day, through Court Records, and Order Book 1678-1698, Part 2, 607, from St. Stephens Parish of Northumberland County, Virginia.

- Mary Day - Generation 1, My 8-G-G-G-G-G-G-G-Grandmother

The following is about Mary Day, (a White Indentured Servant) and Daniel D. Webb, (a Mulatto Indentured Servant). Court Records and other documents seem to indicate that Daniel was the father of at least three of Mary's children. It appears that I can trace my Day ancestors to Mary and Daniel.

From Order Book of 1687, I found that Daniel had petitioned the Court to end his Indenture period; *5-Oct. 1687 - Whereas Daniel D. Webb, a Mulatto son of an English Woman Servant to ye orphans of Major. John Mottrom deceased petitioned this court (having attained the age of one and twenty years) he might be free. It's ordered that if the said Daniel make appeare that he hath attained the age aforesaid and no reasons appears to the contrary from ye honorable Nichol; Spencer who manages the estate of the said orphans, he the said Daniel be immediately free the next Court, Richard Haynie,   Attorney of Daniel D. Webb. OB 1678-98, Part 2, 405.*

Several Court Records were found for Mary Day (a White Woman) who was apprenticed in Northumberland County, Virginia. 22 Nov. 1689 - The Churchwardens of the Parish having presented to the Court Mary Day for having a bastard child, Eliz Gerard for two bastard children, Alice Hudnall, John Haynie Junrs. Servant woman, Rob Seth's servant woman, Thomas Adam's servant woman, John Hughlett Senor's servant woman, John Cole's servant woman for having each of them had bastard children. It's ordered yet ye Sheriff bring each of ye said women to ye next Court to answer for their said default. OB 1678-98.

In the same order book, Mary is in Court with twins 21 Dec 1692 - John M., and Rachell, two Mulatto children twins being born ye 17 February last, begotten basely on the body of Mary Day, servant to John Webb, are bound by the Court to serve the said John Webb or his assigns until they attain the age of thirty years according to ye prescription of ye law in that case provided. OB 1678-98, Part 2, 607

The third time Mary was in Court appears to be several years later, 22 Nov. 1694 - Samuel Webb (alias) Day, a Mulatto child begotten on ye body of Mary Day, is by the Court ordered an Apprentice to serve Wm. Yarratt until he shall attain to age of law prescribed. OB 1678-98. Part 2, 681.

(1)   First Childs name is unknown - born in Northumberland County, VA 1689.

(2)   John Day, I., (twin) - born in Northumberland County, VA 1692

(3)   Rachel Day, I., (twin) - born in Northumberland County, VA 1692

(4) Samuel Day, - born in Southampton County, VA 1693

The following list of activities that were going on during the 1660's, 1670's, and 1680's, shows how very difficult it must have been for Mary and Daniel.

In 1662, Virginia declared that a mother's status determined whether a child was born free or into slavery. By 1670, Virginia had passed a law that declared any person arriving by sea who was not a Christian could be held as a slave for life.

In 1663, a law was enacted in Maryland that said, a free white woman who married a black man was considered a slave as long as her husband was alive. However, a 1681 Maryland law was passed that declared children born to white mothers and black fathers were to be free. The children born to free black women were also to be free.

We have come a long, long way since the early days of slavery in the United States. I really do appreciate the time that I have been able to spend with my Day family members, and am happy that I have learned so much about my ancestors. This history will definitely be passed on to the Day family descendants in Ohio, North Carolina, Virginia, and throughout the world.

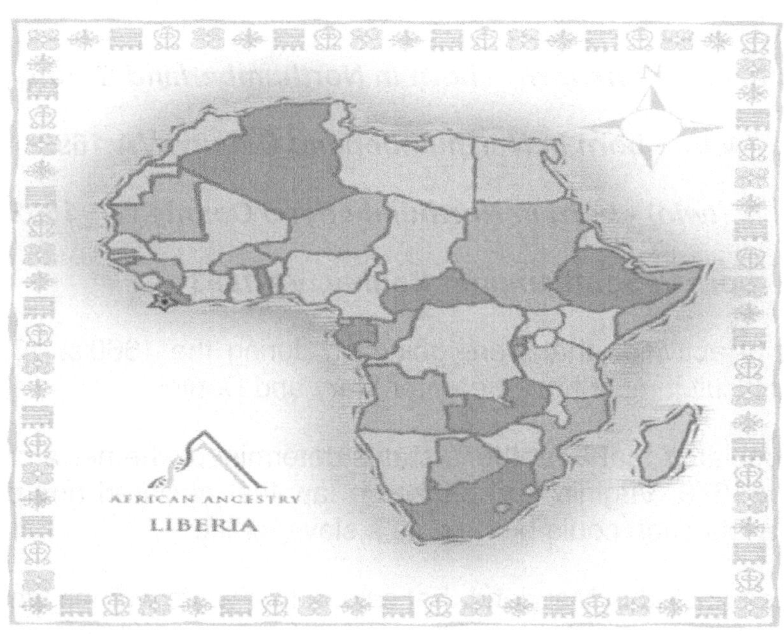

John J. Day, Jr., Superintendent of the Baptist Mission,   a signer of the Declaration of Independence (Liberia), and the second Chief Justice of the Liberia Supreme Court

North Carolina Museum of History.

Top - The statue of Thomas J. Day, Cabinetmaker of Caswell County, North Carolina, is one of three located at the entrance to the museum. (Aaron L. Day on right)

Bottom L-R, Sauratown woman/from early tribe, Cornel Fred A. Olds, & Thomas J. Day

Aaron L. Day visits the Union (Yellow) Tavern in Caswell County, North Carolina.

The roof caught on fire and had to be restored. The Thomas Day Restoration Committee received a $250,000.00 matching Government Grant to renovate the building. It is now a Museum.

James H. Day (Dad's) House in Ohio – (He was the last of his generation)

Top, L-R, Middle, L-R, Frederick Day, Theodore Day, Sharon Day, Aaron L. Day, William

Middle, L-R, Cousins - Pat Day, Ted Day, Jr. Billy Day, Hurrell Bumpas, James Day, III, (Phil), and Luther

Bottom, L-R,   Fred, Ted, Norma Jean, Aaron, William

Day Family in Ohio

Documents for Generations 1-14

If there is a hero to this story I would say that it would have to be my great-great grandfather, Scott Day (Generation 7). Without the numerous documents and records found for him, it would have been almost impossible to trace and connect the Day family back to 1660, and beyond. It was as if I was guided by him back through seven decades of American History, as I learned about his life and times. Scott was almost sixty five years old when the Civil War ended in 1865. Unlike many blacks of the time, Scott had been a free man, all of his life, in the Southern state of North Carolina. He was still considered to be a free man during his long indenture period. Scott and his father Thomas both lived past the age of eighty.

Daniel Webb, generation 1, would have been born approximately 1666, as his petition to the court in 1687 for his release from indenture stated that he had attained the age of twenty-one. There was a Daniel Webb, and Isabel (colored), located in earlier tax records. It is possible that these may have been Daniel's parents, but I have not found records to verify this. Daniel's indenture record stated that he was the son of an English servant woman. Who knows, this may be another long-lost generation, and another story!

(Documents located in Virginia, North Carolina and Ohio)

* **Daniel Webb,** I - 1646 - Isabel

(Father of Daniel Webb - Generation #1) ??????

(1.), Tax records for Gen. John Custis (Curtis), listed are Daniel Webb & Isabel (colored). Not sure, there may be a possible connection!

Generation 1

* **Mary Day - (1671) ---------------------------Daniel   D. Webb, - 1666 - (no marriage record)**
(Mary is the daughter of ??)   ---------------------------(Daniel is the son of ??)

(1.), 1689, 1692, 1693 and 1703 Court Records of births of children - St. Stephens Parish in Northumberland County, Virginia. (2), Additional records for Daniel D. Webb – Indenture Record, and Release from Indenture Record.

Generation 2

* **Rachell Day (alias Webb) - (1692) -------------------- Arthur/Will - (no marriage records)**
(Rachell is the daughter of Mary Day and Daniel D. Webb)

(1.), 1692 Court Record - St. Stephens Parish (shows birth of Rachell and her twin brother John, to a Mary Day. They were listed as Mulattoes (of mixed parentage).

(2), 1709 and 1712, Court Records from Northumberland County, Virginia show the births of her children - she was listed as Rachell Day (alias Webb).

Generation 3

*** Winnifred Day (alias) Webb - (1712) (Winnifred is the daughter of Rachell)**

(1.), 1712/1713 Court Records, Winnifred Day, (Alias. Webb) daughter of Rachell,

(2), 1735, 1740, 1753, birth records of children, (Thomas -1735, John - 1740, George - 1753).

Generation 4

*** Thomas Day, I - (1735) - (Thomas is the son of Winnifred)**

(1), 1735 birth record, his mother was Winnifred, (records from Virginia, and North Carolina).

(2), 1779 & 1789 Caswell County, North Carolina Land Entries.

(3), Days and Winsteads in St. Stephens Parish records, Northumberland County, Virginia, since the 1690's

Generation 5

*** Rachel Day – (1760) - (Rachel is the daughter of Thomas Day, I.) It appears that Rachel was named after her great-grandmother, Rachell.**

(1), A Will for Samuel Winstead, I, was located at St. Stephens Parish, in Virginia - (These records are from Virginia & North Carolina).

(2), Land Entry Deed, Samuel Winstead, III, buys property in Caswell County, North Carolina in 1777, three years before Thomas and George Day were indentured to him. The Days and Winsteads, likely moved together from Virginia to North Carolina between 1777 and 1780. There were Days already living in North Carolina at the time.

(3), 1780 Caswell County, North Carolina Apprentice Bonds show Rachel as mother of Thomas R., and Jessie who were indentured to Samuel Winstead, III). It also shows Rachel's sister Ann, with her children George and Lucy.

(4), Thomas Day, I. - Land Records for 1779 and 1789 show that Thomas was living near Aisley Winstead, the son of Samuel Winstead.

(5), 1783 court record - Rachel's, and Ann's children were listed as orphans

Generation 6

*** Thomas R. Day - (1777) ---------------------- married Millie (per Census Schedules) (Thomas R. Day is the son of Rachel Day). It appears that Thomas may have been named after his grandfather, Thomas Day, I.**

(1.), 1780 Caswell County, North Carolina Apprentice Bonds (Thomas R. Day - 3 years old, son of Rachel, when apprenticed to Samuel Winstead)

(2), 1793 Tax records for Samuel Winstead, Thomas was listed as 16 years old

(3), 1820 & 1850 Census Schedules, (Thomas R. Day, and family)

Generation 7

*** Scott Day - (1803) ------------------------- Married Setta (per Census Schedules), (son of Thomas R. Day)**

(1), 1820, 1830, 1840, 1850, 1860, 1870 Census Schedules. He was located in six Census Schedules.   (Scott was free in North Carolina all of his life).

(2), 1867 Marriage Bond of daughter Milly and Washington Day, (listed Scott and his wife)

Generation 8

*** Milly Day - (1843) --------------------- married Washington Day (per Marriage Bond and Census Schedules) (Milly is the daughter of Scott Day). It appears that Milly may have been named after her grandmother, Millie Day.**

(1.), 1870 & 1880 Census Schedules - Milliy is listed with children

(2.), 1867 Marriage Bond with husband Washington Day - shows that Milly is the daughter of Scott and Latta (Setta) Day

Generation 9

*** James L. Day - (1862) -- married Sallie Bumpas (Bumpers), (per Census Schedules)**
(James is the son of Milly Day)

(1.), 1880, Census Schedule. (James was listed on the 1880 Census as 19 years of age, with his cousin Richmond Day, and his family)

(2),   He was also on the 1910 and 1920 Census Schedules with his wife Sallie, and their children.

Generation 10

***   James H. Day - (1907) ----- married Helen Graves/Lula Banks/Lena Stanford (Marriage Records) (James is the son of James L. Day)**

(1.), 1910 & 1920 Census Schedules. James was listed on the 1910 and 1920 Census Schedules with his mother, father, sisters, and brothers.

(2), Family Documents.   (Records listed here are from North Carolina and Ohio).

Generation 11 -   OLDEST LIVING GENERATION

(children of James H. Day)

*   **William Day -** married Beverly Pullen/Kathy Jones

*   **Shirley Ann Day -** deceased in childhood

*   **James Day, Jr.,** (deceased) - married Shirley Freeman/Wadie Meaux

*   **Aaron L. Day - (1939) -** married Dorothy Lindsay

*   **Theodore Day -** married Carolyn Fairfax

*   **Frederick Day -** married Rosemary James

*   **Sharon Lou Day -** married Doug Thompson

(1.), Birth Certificates/Marriage Records

(2), Family Documents, (all records here and below from Ohio).

Generation 12

***   Carla Day-Moody-Harris - (1961) ------------married Clois Ray Moody/Jerry Harris (Carla is the daughter of Aaron L. Day)**

(1.), Birth Certificates

(2), Family Documents.

Generation 13

SIBLINGS - (children of Carla)

* **Crystal Moody-Harris -** (1983) --------------------- **married Sean Anthony Harris**

* **Ray Moody - (1985)**

* **Ephrom Moody - (1989)**

* **Arrionne Moody - (1993)**

* **Maliah Harris - (1999)**

(1.), Birth Certificates

(2), Family Documents.

Generation 14

SIBLINGS - (children of Crystal)

* **A'Sean D'Anthony Harris** - (2003) & **Tyray** (2007),

(1.), Birth Certificates

(2), Family Documents.

There have been many changes through the different generations. **On May 17, 1954** the United States Supreme Court ruled in Brown vs. Board of Education, Topeka, Kansas, that racial segregation in the public schools was unconstitutional. Although I was only fourteen years old at the time, I still remember the excitement generated in my community by the Brown vs. Board decision. My teachers kept all of the students up-to-date on the proceedings. "By Invitation Only" tells of one notable experience that happened during that time period.

By Invitation Only

How many hues of color are there? How do these colors affect us as individuals? There are black, brown, white, yellow, red, and many more shades of people. At what age are we when we learn that color may have a significant influence on our lives? Like learning to walk, talk, read, and write, we will each discover the meaning of different colors at various stages of our lives.

For me, it was in 1956 at the age of sixteen, shortly after I received my first driver's license. I learned more about color that year than I had during the previous fifteen years of my life.

With my newly acquired driver's license, I set out to drive to a movie theatre about twenty miles away. How exciting it was for me to be able to drive alone for the very first time. All of those months spent learning to drive had finally paid off. It would be nightfall when I left the movies to return home. Driving in the dark would be another new experience for me, and again, my driving skills would be tested.

I was looking forward to seeing the new movie, which I had heard so much about. In those days, movies were my passion. I could name all the academy award winners of the past, and even tell you what movies had won for best picture of a particular year.

As I walked up to the cashier at the movies, my excitement was suddenly crushed, by what I heard. "Sorry, but this theatre is by invitation only." By invitation only, I repeated? It was then, that she slightly lowered her head and said, "I'm very sorry."

In that brief moment, I suddenly began to understand all of the unexplained instances of the past that had always puzzled me. The restaurants that we never ate in, the neighborhoods that we never lived in, the churches that we never joined, and the schools that we never attended, was because of the color of our skin. Although, there were no signs – white only, or no colored allowed, it was understood. I now clearly understood the meaning of "By Invitation Only."

Fortunately, there are many people like the cashier, who are truly sorry for the unequal treatment of others. Through the years, in some areas things have gradually changed for the better, and continue to improve. In other areas, the struggle continues.

By Aaron L. Day

PART II

DNA TO AFRICA

And

THE DISCOVERY OF MY HOMELAND

Locating Free African American Ancestors

Book by Aaron L. Day

CHAPTER 8

DNA TO AFRICA

&

THE MOLECULAR GENEALOGY
RESEARCH PROJECT - 2001

DNA to Africa

MOLECULAR GENEALOGY RESEARCH PROJECT

It was in 2001 that I first heard of the link between DNA and genealogical research. The Sorenson Foundation and The Church of Jesus Christ of Latter-day Saints were working together to establish a large DNA database that could eventually be used for genealogical research. It was called, "The Molecular Genealogy Research Project." I was very intrigued by the information on the flyer advertising this DNA Project. Could this DNA Project possibly help me with my genealogical research? The flyer stated that they were having a conference in Carlsbad, California to give details about Molecular Genealogy and also to collect blood from individuals who wanted to participate in the project. The North San Diego County Genealogical Society sponsored the event.

A friend of mine and I decided to drive from Long Beach and spend the weekend in San Diego to do some sightseeing and also to participate in the conference. The conference in Carlsbad on November 17, 2001, turned out to be very exciting and enlightening. I took a seven-generation family pedigree/ancestor chart with me and donated a sample of blood to be added to the DNA Project database. A major portion of this research is to identify and trace ancestral origins.

It was a great weekend, and I really did enjoy myself. I made many new friends and established connections with numerous professional genealogists at the conference. I began to learn more about the Molecular Genealogy Research Project. Molecular Genealogy links individuals together in 'family trees' based upon the unique identification of genetic markers.

The information encoded in the DNA of an individual and/or population is used to determine the relationships of individuals, families, tribal groups and populations. DNA may be obtained from a person by sources such as blood, saliva, and hair.

The major objectives of the Molecular Genealogy Research Project are to:

* **Determine the genetic composition of major populations throughout the world.**

* **Reconstruct genealogies using genetic information.**

* **Establish genotypic links (screening for specific genetic markers in the laboratory) in each population and between each of the populations.**

* **Produce unique identifications for persons who do not have traditional name-based genealogies.**

* **Preserve the genetic heritage of an individual and family for future generations.**

The Sorenson Foundation continues to conduct research into new methods of using DNA analysis for genealogical research. Over the past few years a number of other DNA research and testing projects have appeared. These companies are researching, testing, and establishing databases that may be used for genealogical purposes. Anyone interested in learning more about DNA and the link to genealogical research may be able to:

* **Determine if two people are related**

* **Confirm your family tree**

* **Determine if two people descended from a common ancestor**

* **Find out who with your surname is related**

* **Prove or disprove a research theory**

* **Find and confirm new individuals in your family tree**

* **Check paternal ancestry**

* **Check maternal ancestry**

* **Estimate ancestry (percentage)**

* **Check for paternity**

* **Learn more about - surname**

* **Check ethnicity**

* **Determine origins such as:**

* **African ancestry**

* **East Asian ancestry**

* **European ancestry**

* **Jewish ancestry**

* **Native ancestry, and more**

In 2004, I participated in the 4th annual "West Coast Summit on African American Genealogy" in San Diego. I gave a workshop on "Searching for free African American Ancestors pre-1865." The workshop was about the 'Day' side of my family. I had discovered that many of my Day ancestors were free long before the Civil War ended in 1865, and wanted to share this. Also making a presentation at the Summit was Gina Paige, President of the African Ancestry Company Incorporated. Unfortunately, I was unable to attend her workshop, as I had one scheduled at the same time, but I picked up literature about her organization and learned more about Y-DNA and Mitochondria DNA and genealogy.

The African Ancestry literature explained why tracing ancestry might be important to some people. It also explained how the DNA testing process works, and the accuracy of the process, **(See Appendix II).** I was very curious and wanted to learn how some people might use their results. This was also explained in the African Ancestry literature. From the literature, I learned that some people use it:

* **To augment their family tree and share at Family Reunions.**

* **To further personal, family, cultural, and historical education.**

* **To connect with African communities within the United States**

* **To support political causes on behalf of African Nations.**

* **To travel to their ancestral homeland.**

In February of 2005, I heard that African Ancestry Inc. was giving a presentation on DNA/Genealogy at the Pan African Festival in Los Angeles, California, and decided that I had to attend. I was not let down, as Professor Rick Kittles gave an excellent presentation on DNA and genealogical research. Dr. Kittles has worked with the National Human Genome Center at Howard University in Washington, D.C. Dr. Kittles has built a database of over 22,000 DNA samples representing 135 African population groups. Dr. Kittles, and Gina Paige, formed the pioneering African Ancestry Company Incorporated. In his presentation Dr. Kittles talked about:

* **The Y Chromosome which is used to determine paternal ancestry**

* **How the Y Chromosome is read**

* **The Mitochondria DNA (mtDNA) which is used to determine maternal ancestry**

* **How the mtDNA is read**

* **How the Y Chromosome and mtDNA are analyzed**

* **How the results of the tests are determined**

Professor Kittles' presentation convinced me that I should order the MatriClan and PatriClan kits so that I might learn more about my maternal and paternal African ancestry. After receiving the kits in the mail about a week later, I sent back a sample of my DNA. Unlike the previous project in Carlsbad five years earlier where a blood sample was required, I simply sent a DNA sample by taking a swab within my mouth and rubbing against my cheeks. This sample was put in a small envelope and mailed to the African Ancestry laboratories. In about six weeks I received a packet from African Ancestry with the results.

The first paragraph of the letter I received from African Ancestry revealed what I had been so anxious to learn. "It is with pleasure that I report our MatriClan analysis successfully identified your maternal genetic ancestry and our PatriClan analysis successfully identified your paternal genetic ancestry. The mitochondria DNA (mtDNA) sequence that we determined from your sample shares ancestry with the Hausa people in Nigeria. The Y-chromosome DNA sequence that we determined from your sample shares ancestry with the KRU people in Liberia." The report also included details on how my African ancestry was determined.

I had finally traced my bloodline to Africa. It was a wonderful feeling to learn about my African ancestry. As you can imagine, I could hardly contain my enthusiasm that day. I wanted to tell everyone. I called my siblings, cousins, aunts and uncles and told them about the results. They were all very happy to hear the good news.

I remember when I told my daughter Carla about it she said "I want to learn more about the history of our family so that I will know what to tell the children. They will also be able to tell their children about our family history."

This is what has happened so far

A male passes Y-DNA on to his sons, but not to his daughters. I received the Y-DNA from my father, James H. Day, who received it from his father James L. Day, and so on back through the male line. This Y-DNA was tested back to the KRU people of Liberia. However, I did not receive mtDNA from my father's mother. On my paternal side I have a female cousin who received mtDNA from her mother and she has volunteered to send in DNA. We should then be able to determine the ancestry of my paternal grandmother.

A female passes mtDNA on to all of her children, however, only the daughters pass it on to their children. I received the mtDNA from my mother Lula Banks Day, who received it from her mother Olla Bondurant Banks, and so on back through the female line. This mtDNA tested back to the Hausa people of Nigeria. However, I did not receive Y-DNA from my mother's father. On my maternal side I have a male cousin, Charles Banks, who received Y-DNA from his father and he volunteered to send in DNA. The results revealed that my maternal grandfather, George H. Banks, descended from the IBO people of Nigeria.

One of my cousins called and told me that her son had given her a DNA Test Kit for a Christmas present. She wanted to know which DNA Test she should ask for. Her mother, Emma Banks-Thompson-Thomas, and my mother, Lula Banks-Day, were sisters, so if she did the mitochondria DNA test it would probably lead to the Hausa people of Nigeria as mine did. I suggested that she take the DNA test that would estimate ancestry percentage among the four major population groups. Her results were:

Sub-Saharan African – 59%

European or White – 33%

Native American – 0%

East Asian – 8%

Oral history in our family led us to believe that our maternal grandmother, Olla Bondurant Banks was part Cherokee. We grew up hearing this story all of our lives. Mother and daughters had an olive skin coloring, with thick, coarse, wavy, beautiful hair. This was part of our heritage, so we thought the DNA test would finally verify that part of our family history.

The following chart is for the 'Day' side of the family. This chart includes my family and also the family of my cousin Dora Day-Parks (generation #11). There is an unusual twist to this chart. From generation #11 to generation #8 the Y-DNA traces back to Washington, and then veers off of this particular chart. The Liberia ancestry is through Washington, his father, grandfather, and so on back through the male line.

Generation 1   -   1671 - Mary Day-----Daniel D. Webb - 1666

Generation 2   -   1692 - Rachell Day (Webb) ----- Arthur/Will

Generation 3 - 1712 - **Winnifred Day (Webb)**

Generation 4 - 1735 - **Thomas Day, I**

Generation 5 - 1760 - **Rachel Day**

Generation 6 - 1777 - **Thomas R. Day, II ----- Millie**

Generation 7 - 1803 - **Scott Day ----- Setta**

Generation 8 - 1843 - **Milly ----- Washington Day**

Generation 9 - 1862 - **James L. Day ----- Sallie Bumpas**

Generation 10 - 1894 - **Wilborn Day ----- Ila**
                                    (Wilborn & James - Brothers)

Generation 10 - 1907 - **James H. Day ----- Helen/Lula/Lena**

Generation 11 - 1935 - **Dora Day-Parks ----- James Parks, Sr.**
                                    (Dora & Aaron - Cousins)

Generation 11 - 1939 - **Aaron L. Day ----- Dorothy**

Generation 12 - 1959 - **James Parks, Jr.**

Generation 12 - 1960 - **Kevin Parks**

Generation 12 - 1961 - **Carla Day-Moody-Harris ----- Ray/Jerry**

Generation 13 - 1980 - **Jennifer Parks**

Generation 13 - 1981 - **Terrell Parks**

Generation 13 - 1982 - **Darnell Parks**

Generation 13 - 1983 - **Crystal Moody ----- Sean Anthony Harris**

Generation 13 - 1985 - **Ray Moody, Jr.**

Generation 13 - 1989 - **Ephrom Moody**

Generation 13 - 1993 - **Arrionne Moody**

Generation 13 - 1999 - **Maliah Harris**

Generation 14 - 2000 - **Jahler Parks**

Generation 14 - 2003 - **A'Sean D'Anthony Harris & Tyray Harris (Crystal's sons)**

I began learning more about other DNA testing companies and the different services that they provide. The following is a listing of a few of the companies that are available, but is not a complete list.

*   African Ancestry Company – (African ancestry)
www.africanancestry.com

*   Ancestry by DNA – (Native American, European, East Asian, African, & estimated ethnicity)
www.ancestrybydna.com

*   DNA Heritage – (Y-DNA)
    www.dnaheritage.com

*   DNA Print Genomics – (Native American – examines main body of DNA & gives percentage of ethnicity)
    www.dnaprint.com

*   Family Tree DNA – (surname, Y-DNA, mtDNA, Native American, Jewish, DNA Plus, geographic origin, African American, Western European origin, +)
    www.familytreedna.com

 * AfricanDNA.com
    www.africandna.com

 * Gene Tree – (DNA Paternity, +)
    www.genetree.com

*   GeoGene – (Y-DNA, mtDNA)
    www.geogene.com

*   Oxford Ancestors – (Y-Line –surname project)
    www.oxfordancestors.com

*   Relative Genetics – (Y-chromosome, mtDNA, ethnicity)
    www.relativegenetics.com

*   Roots for Real – (mtDNA)
    www.rootsforreal.com

*   Sorenson Foundation– (Building a DNA database, conduct research, +)
    www.smgf.org

*   Trace Genetics LLC – (mtDNA, Y-DNA, specializing – Native American)
    www.tracegenetics.com

Another very interesting book, is called   "Trace Your Roots with DNA." This book is by two pioneers in the field of Genetic (Genetealogy) Genealogy, Megan Smolenyak Smolenyak and Ann Turner. www.rodalestore.com

Book recommendations:

*   **DNA & Genealogy,** by Colleen Fitzpatrick and Andrew Yeiser, www.forensicgenealogy.info

*   **Forensic Genealogy,** by Colleen Fitzpatrick, www.forensicgenealogy.info

*   **Trace Your Roots With DNA,** by Megan Smolenyak Smolenyak, and Ann Turner, www.rodalestore.com (2004) - ISBN 1594860068

*   **DNA & Family History,** by Chris Pomery, www.DNAandfamilyhistory.com

Additional recommended readings

*   **Priscilla's Homecoming: A Remarkable Journey,** by Mrs. Thomalind Martin Polite, http://www.africanaheritage.com

*   **USA Today article in Science and Space,** DNA re-writes history for African Americans, By Richard Willing, posted Feb. 01, 2006, http://www.usatoday.com

*   **The Miami Herald, Journey Yields Glimpse of Past,** by Paul Simon, posted on Sun, Nov. 07, 2004, http://www.herald.com

*   **Everton's Genealogical magazine, DNA to Africa,** by Aaron L. Day, published July, 2006, http://www.day-banks.com & http://www.scgsgenealogy.com

*   **Tracing your Ancestors, featured in the Black Enterprise magazine,** by Allison Gilbert, Aug., 2005, http://www.africanancestry.com

PBS - DNA Documentary

*   **African American Lives,** was a four-hour documentary series on Genealogy/DNA. The show focused on eight well-known African Americans and the process of using genealogy and DNA to locate their African homelands. Henry Louis Gates, Jr., hosted the program during the month of February in 2006.

Henry Louis Gates., Jr., is the author of 'Finding Oprah's Roots.' He recently established a DNA testing company project with, Family Tree DNA.

AFRICAN ANCESTRY EDUCATION PROJECT

African Ancestry continues to break new grounds in the field of DNA and Genealogical Research. In January, 2009, African Ancestry teamed up with the **Walker Jones Education Campus in Washington, DC** to celebrate Black History Month, and trace the roots of 10 very deserving and excited 7th graders. African Ancestry went to the predominantly African American school to collect DNA samples from Principal Jeffrey Grant, the World History and Life Sciences teachers, and the students, as well as educate the class on the power of DNA and the great continent of Africa.

In February, African Ancestry returned to Walker Jones to reveal their ancestry and celebrate a new part of their identity with their family and friends. The students learned more about ancestry, identity, and Africa through an interdisciplinary curriculum provided by African Ancestry. The Ancestry Education Project had a powerful celebration, where African Ancestry assembled the students, their families, faculty, and the community to unveil where in Africa, each student's bloodline originated.

I just learned that the 100 Black Men of Greater Washington, D.C. is partnering with African Ancestry. President, Marvin Dickerson tells why it is important for the young students they mentor to find their roots. The DNA results are presented to the young African American men so that they can learn more about family and heritage. I will be conducting classes for The 100 Black Men of Long Beach, Inc., and helping the young students to fill out Pedigree Charts and Family Group Sheets. There may be a possibility that our Chapter will also work with African Ancestry, and introduce the students to the DNA/Genealogical connection.

I am so happy that I heard about the link between DNA and genealogical research. It has been great to learn so much about the Sorenson Foundation, the Church of Jesus Christ of Latter-Day Saints, the Molecular Genealogy Research Project, and the ever-growing DNA testing sites and projects. The DNA to Africa connection has been very exciting. I am looking forward to learning more about the lives of my free ancestors, and their DNA to Africa connection.

With my DNA tests, and the discovery that my parents' ancestors were from Liberia and Nigeria, I have been learning about the history of West Africa. I also began meeting people from other parts of Africa, the North, the East, the South, and the Central areas. Within the past few years, I have made many new friends who are from Africa.

My traditional genealogical research through census schedules, family papers, church records, land deeds, free register papers, court documents, and other records has been very productive for me. I have traced my family back to 1666. These records were located in the States of North Carolina, and Virginia, and now I have included - Africa. Hopefully, the details about DNA and family history research that I have outlined here will help you in your future research and documentation of your family history. For anyone wanting to learn more, there are excellent books on DNA and using genetic tests to explore your family history.

SPECIAL PROGRAM

THE MOLECULAR GENEALOGY RESEARCH PROJECT
OF
BRIGHAM YOUNG UNIVERSITY

Saturday
17 November, 2001
10AM to 2 PM
Church of Jesus Christ of Latter-Day Saints
1981 Chestnut Avenue
Carlsbad, California

Sponsored by the North San Diego County Genealogical Society
And the Family History Center, Carlsbad
Presentation at 10 AM

Blood sample drawing for those wishing to submit a 4-generation
Pedigree chart
11:15AM to approximately 2 PM
Must be 18 years or older to participate

Consent form to participate will be available at a later date
And at the Program
4-generation pedigree chart on reverse
Please RSVP by Nov. 1, 2001 The first 300 reservations will be taken

- -

Yes! I want to participate in this exciting study on November 17th!
Name_______________________________________
Address____ Phone____________
E-mail Address_________________________________
Mail to NSDCGS, P. O. Box 581, Carlsbad, CA 92018-0581 or
E-mail information to fwspong@CTS.com
First 300 reservations will be accepted so don't delay!

Molecular Genealogy Flyer, 2001

AFRICAN ANCESTRY

Aaron L. Day

Hypervariable Segment I of mtDNA (HVSI starting at position 16024):

```
TTCTTTCATG GGGAAGCAGA TTTGGGTACC ACCCAAGTAT TGACTCACCC ATCAACAACC
GCTATGTATC TCGTACATTA CTGCCAGCCA CCATGAATAT TGTACGGTAC CATAAATACT
TGACCACCTG TAGTACATAA AAACCCAATC CACATCAAAA CCCCCCCCTC ATGCTTACAA
GCAAGTACAG CAATCAACCT TCAACTATCA CACATCAACT GCAACTCCAA AGCCACCCCT
CACCCACTAG GATATCAACA AACCTACCCA TCCTTAACAG TACATAGTAC ATAAAGCCAT
TTACCGTACA TAGCACATTA CAGTCAAATC CCTTCTCGCC CCCATGGATG ACCCCCCTCA
```

Sequence Similarity Measure: 99.7%

Your sequence is 99.7% the same as sequences from Hausa people in Nigeria.

The bold letters indicate DNA sequence patterns that you share with the Hausa.

Y Chromosome (NRY) Polymorphisms:

MARKER	DYS388	DYS389I	DYS389II	DYS390	DYS391	DYS392	DYS393	DYS394	YAP
ALLELE SIZE	12	13	30	21	10	11	13	17	+

Sequence Similarity Measure: 100%

Your sequence is 100% the same as sequences from Kru people in Liberia.

African Ancestry 5505 Connecticut Avenue, NW Suite 297 Washington, DC 20015 TEL 202.723.0900 FAX 202.318.0742

African Ancestry - (Nigeria & Liberia)

CERTIFICATE OF ANCESTRY

African Ancestry hereby certifies that

Aaron L. Day

shares Maternal Genetic Ancestry with

the Hausa people in Nigeria

Based on a MatriClan ™
analysis performed on

March 31, 2005

Rick Kittles, Ph.D.
Scientific Director

Certificate from African Ancestry

Hausa people in Nigeria

CERTIFICATE OF ANCESTRY

African Ancestry hereby certifies that

Aaron L. Day

Shares Paternal Genetic Ancestry with

the Kru people in Liberia

Based on a PatriClan ™
analysis performed on

March 31, 2005

Rick Kittles, Ph.D.
Scientific Director

Certificate from African Ancestry

Kru people in Liberia

Charles R. Banks, Jr.

Y Chromosome polymorphisms

MARKER	DYS388	DYS389I	DYS389II	DYS390	DYS391	DYS392	DYS393	DYS394	YAP
ALLELE SIZE	12	13	29	22	10	11	14	15	+

Sequence Similarity Measure: 99.7%

Your sequence is 99.7% the same as sequences from Ibo people in Nigeria.

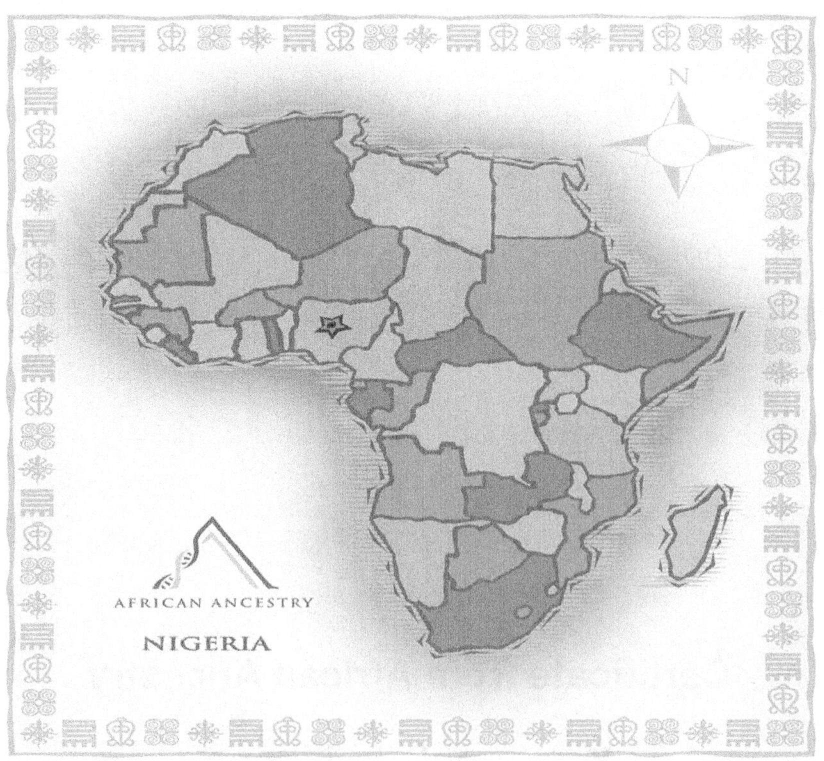

Charles R. Banks, Jr., DNA test results
The Ibo people in Nigeria

CHAPTER 9

THE DISCOVERY OF
MY HOMELAND

The Discovery of my Homeland

After learning that some of my father's ancestors had been free before 1865, I began discovering more about the free people of color. Learning about the DNA and genealogical family history connection has also been a great learning experience. I have enjoyed learning so much about the different countries of Africa, and was happy to make the connection to my homeland. As a result, my research took me back to a very painful period in America - the institution of slavery, and the middle passage. The shipment of slaves from Africa to America was known as part of the Middle Passage.

Slaves were carried by ships from Africa through the West Indies to North America. Trade goods would also be shipped from North America to Africa. Captured slaves were roped and shackled together on ships from Africa to North America that sometimes took many weeks to cross the ocean.

I began to think about my early history classes. It seems like each generation is re-writing history. I distinctly remember learning about the slave trade from my teachers. I learned that some African Chieftains did sell their people into slavery, but that was not the only way our people became slaves. Today, our young people seem to be learning that the only reasons Blacks became slaves, was because their Chiefs sold them into slavery. However, we know this is not the complete truth, and need to stop that myth. It is the same as saying 'the Chiefs let the different countries come and colonize, or take over Africa.'

I grew up reading the National Geographic magazines, and enjoyed learning about other parts of the world. The maps about the different countries in Africa were especially meaningful for me. Geography was always one of my favorite subjects. I also found the history of the United States interesting, and at times, very puzzling. The institution of Slavery was very upsetting for me to deal with.

All of the 13 original colonies legally recognized slavery.

*** Massachusetts - 1641**
*** Connecticut - 1650**
*** Virginia - 1661**
*** Maryland - 1663**
*** New York - 1665**
*** South Carolina - 1682**
*** Pennsylvania - 1700**
*** New Jersey - 1702**
*** Rhode Island - 1703**
*** New Hampshire - 1714**
*** North Carolina - 1715**
*** Delaware - 1721**
*** Georgia - 1755**

Slavery in America did much to disconnect slaves and free Blacks from their African heritage. The institution of slavery has created a wall of separation between the Blacks in Africa, and the Blacks in America. Most people living in America know their connections to other countries, and many maintain a relationship with family and friends. However, due to slavery, Blacks do not have that connection. We must bust through that wall, go over it, or around it, and re-establish our connections to Africa.

While slaves were being brought to America, the Continent of Africa was being taken over (colonized) by other Countries like Great Britain, Egypt, Italy, France, Portugal, Belgium, and Spain.   The years the different countries gained their independence, are listed below.

1931 - South Africa	1961 - Sierra Leone
1956 - Sudan	1962 - Uganda
1957 - Ghana	1962 - Burundi
1958 - Guinea	1962 - Rwanda
1960 - Senegal	1963 - Kenya
1960 - Mauritania	1964 - Zambia
1960 - Mali	1964 - Tanzania
1960 - Ivory Coast	1964 - Malawi
1960 - Upper Volta	1965 - Rhodesia
1960 - Togo	1965 - Gambia
1960 - Benin	1966 - Botswana
1960 - Nigeria	1968 - Equatorial Guinea
1960 - Niger	1968 - Lesotho
1960 - Cameroon	1968 - Swaziland
1960 - Chad	1974 - Guinea-Bissau
1960 - Central African Empire	1975 - Comoru Islands
1960 - Gabon	1975 - Angola
1960 - Congo	1975 - Mozambique
1960 - Zaire	1977 - Djibouti
1960 - Somolia	
1960 - Malagasy Republic Madagascar	

Many of my new friends from Africa are now living throughout the United States of America, and periodically return home to Africa for visits. I enjoy hearing them talk about the different countries of Africa and learning more about the various cultures. Likewise, they enjoy hearing me talk about my life here in the United States. One question I ask my friends from Africa is; **what would you like to tell African Americans about your Country?** The following is the information about Africa that my friends wanted me to include in my book. My friend Issakson Nabila from Ghana sent me an e-mail, and his wife sent me his picture.

My friend Issakson from Ghana

Sir, this is Issakson Nabila, your friend you met at the library. Below is just a brief comment on Africa, but if there is any other specific area that you wish to know about, just let me know.

Africa as you know is a land of many diverse cultures and resources. There is however, a lack of infrastructure and economic development due to the colonial exploitative policies which the continent is still struggling to overcome. Africa is a poor country (as it is made to believe). Africa's woes today were carefully architectured by the colonial masters for their industrial development, long before their exit. The colonial masters did not willingly leave Africa for the Africans to take up their destiny into their own hands very easy. So, they left the continent without constituting any blueprint or plan for its economical, industrial and most of all, political development. They were only interested in the extraction of the continents resources to feed their industries. So, there were no political and economical developmental directions for the freedom fighters who took over, to follow. Besides, they instigated other politicians to stand up against their leaders and eventually things went overboard and resulted in civil disobedience, then into civil wars.

Africans or Black people I should say are very loving and kind people and the colonial masters realized this, so they carefully took advantage of these characteristics and exploited them. Africa is the only place in the world where a total stranger would be accepted into a family or home without any prejudices. Today, they are paying a very painful price for their humility, kindness and hospitality. Africans had their own civilizations before the arrival of the intruders, who first came as missionaries and later became masters or rulers and enslaved the Africans and exploited their resources. Today, we hear the same colonial masters organizing summits and conferences to prescribe an antidote for African woes. These prescriptions are not really genuine prescriptions for the continents woes, but rather to increase the dependability of the continent on them. It is not in the interest of the masters to see Africa become totally economically independent of a competitive force to reckon with.

My friend Cynthia from Cameroon

Africa is not like you African Americans think it is. It is very sad that many African Americans don't know much about Africa. I enjoy living in, and meeting people from the United States. I am a Registered Nurse and I am planning to continue my education and get a Masters Degree. My parents live in Cameroon, and are both highly educated.

There are many highly educated Africans in Africa, especially in my country, Cameroon. Unlike what you hear about us in the United States, there are many highly educated people in Africa. There are also many schools, houses, businesses, shops, churches, etc.

Just remember, that all Africans are not lazy and shiftless, and we definitely don't live in trees. I am learning a lot about the United States of America, and wish that African Americans can also learn about people and life in beautiful Africa. I hope that African Americans will continue to make sure that their young people continue their schooling, and learn more about Africa. Stop watching the movies, and take a wonderful vacation to Africa, and see for yourself.

My friend Edward from Ghana

I believe that for the past 400 years or so, the struggle in this land, the American Continent has been very difficult for Black people. I think that Black people should remember the land of their heritage. All Black people should be acknowledged of the fact that they have a peaceful continent of abode (AFRICA). They should remember from the past to the present, that there is a time for everything in life. There is a time to mourn and a time to enjoy.

The 400 years of human atrocities, mischievous conditions and immoral afflictions are yet to be over. The sick, the afflicted, the tortured, and all can be sent back home to Africa. No one should be left behind when the time comes for departure. Forget about the wealth of this land. Your ancestors helped to bring this country into existence, but were never given their fair share. You are wealthier than any race on this planet when you go back to Africa. Forget the past tribulations, and educate your children well, for the future lies in them. Tell them the history of Africa and America, and all of your ancestors. Think of going to Africa - the home of your ancestors.

My friend Peter from Uganda

Hello Aaron, this is your friend Peter. This is what I want to tell Americans about Uganda, and Africa. Uganda, like many other African countries, is a beautiful place, but because of its distinct features (natural resources and geographical features, it's popularly known as the **Pearl of Africa.** It's found on the Eastern part of Africa and is ecologically located where the East African Savannah meets the West African jungle. It is among the least places in the world today where one can observe lions prowling the open plains in the morning and track chimpanzees the rainy forests undergrowth, the same afternoon, then navigate tropical channels teeming with hippos and crocodiles. With its presence on the Equator, it enjoys a wonderful climate throughout the year, with four different seasons, all spread out. All in all, one can say that Uganda enjoys an extended summer season into the winter. This is good for the travelers, and also the indigenous people.

Uganda has a diversity of cultures and traditions. This small country stretches with more than 50 different tribes and each with different cultural norms and traditions, but all very hospitable. Because of this tradition of hospitality, this beautiful country is best known for being **Africa's Friendliest Country**. It has an extremely low level of crime rate, and hostility directed at tourists and visitors. With a 40 km's distance to the airport lies Kampala, the capital city of this country. It is sprawled across seven hills. The bright modern feel of this cosmopolitan and bustling city welcomes you to Kampala and reflects the ongoing economic growth and political stability that Ugandans have enjoyed since 1986.

Along the way from the airport to Kampala, one is amazed by the natural sloping, spaciousness, and run-away greenery of the garden setting. Then on the other side, is the beautiful view of the biggest lake in the world, Lake Victoria. This country is so rich with a lot of wonderful and beautiful travel destinations, with six natural National Game Parks; its range of forest primaries is very impressive and almost compared to none. There are plain antelopes, and a variety of beautiful birds singing in the sky. The mighty River Nile has its source in this small country, punctuated by a number of waterfalls. With the snow capped peaks of the Rwenzon Mountains and the volcanoes of mountain Elgon, it offers high opportunities for people who want to go hiking through scintillating highland scenery

Yet, there is more to the country, wild life, sceneries and attraction centers far beyond your imagination. With its riches in natural resources still untapped - the country offers a variety of business opportunities. This is the place for entrepreneurs who are looking to make investments, and live a life of no regrets. This country offers very fertile soils and the biggest economic activity is agriculture. Some of its cash crops include; coffee, tea, bamboo, vanilla, shea nuts, cotton, etc. Do you want to save souls back into the Kingdom? There is no better place than this country. With an overwhelmingly big number of churches springing up throughout, the people are always ready for any person to come to share with them - the word of truth. Because of Christianity and religion, the country has seen the spread of HIV/AIDS go down, and has been honored by the United Nations for its great work. Are you looking for a place to go? I don't think there is any place else that you will enjoy yourself as much. See you there - when you come for your visit!

My friend Mbaiceesay from Nigeria

Hi Mr. Day,

How are you doing today? Sorry that it has taken me almost a week to email you. If you remember, we met at the library last week at the closing time. Sure nice talking to you and getting some of that knowledge of yours.

To answer your question or to fulfill your assignment as what advice I have for Blacks in America, coming from a Black man from Africa. Well, my first advice as to what Blacks here need to know is to first go to the primary sources of everything and inquire about things that you are told. When you go to the primary sources then you can think for yourself instead of someone else thinking for you.

First, let me start with religion. In general, Blacks have let some of their so called religious leaders misguide them on everything religious. Perhaps as a form of financial control, but the consequences of religious failure are very drastic. I wish all the Blacks in this country knew la e laha e lallah, Muhammad Rasullulah. That is to say that, there is no deity worthy of worship except Allah and Muhammad is his messenger. Whoever believes that is a Muslim, and a Muslim means the one who submits his or her will to the one God. Malcolm X once said that if America discovers true Islam, then maybe she can cure the cancer of racism that has gone malignant in her body.

On the economic or the financial front, Blacks need to realize that there is enough wealth in Africa available for them to go get their share. Instead, they listen to people who paint a bleak picture about Africa, yet sets shop to rape the natural resources of Africa. On education, especially history, we listen too much about his story instead of our story. That's why; I think your work on the Heritage of African Americans in Long Beach is a good read for all the Blacks who live in the Los Angeles area. Of course there is more to be said, I just want to stop here for now and hear from you. Feel free to comment or ask questions. I look to hear from you soon.

My friend Peter Lobe, from Cameroon

We have reached a moment in our journey where we no longer face express trials and tribulations. We have conquered all odds. However, to contend that we have knowledge of our destination is a fanciful supposition. All we are, is a group of people, with knowledge of a home, that we were involuntarily compelled to reject.

Although, I do not expect the same magnitude of whips and chains to bondage the slaves' progeny back home to Africa, you – once again must cross the oceans. It is worth noting, that tens of thousands died in an attempt to reunite with their motherland – Africa. They broke the chains, broke the cells, and ran hundreds of miles, only to be caught and bound into chains, again and again.

You are a child born – free, thanks to the relentless uprisings of our predecessors. All they wanted – was to see home again. Unfortunately, they never had a chance to see home, or their loved ones again. Don't we owe them that homage? Go home, reunite, learn the culture, the language, and most especially, learn about the people, and the continent. This time, the trip will be without whips, chains, and cells.

Wouldn't you like to see what it was that your great-great-great-great-grandparents missed so much, that they would choose to die for? To mark what went down in history as the greatest human mistreatment, and the subsequent victory, this is what needs to happen.

Each African, wherever you are, you need at least one visit to Africa – as pilgrimage for the souls that were sacrificed.

I am very happy that I met Peter Lobe, III, and that we have become very good friends. He was very excited when I asked him to contribute information about his country, and that I would include it in this book. He did an excellent job of sharing his views, and it has been great learning about Cameroon.

Peter was born at Ekondo Titi, in the South West Province of Cameroon. What made him want to come to the United States? It was during his early years, that he heard, and learned so much about the United States. When he became an adult, he felt that a trip to the United States would be very interesting.

After obtaining employment in the United States, Peter began attending the Southwestern University School of Law. He is pursuing a Masters' Degree in Law (LLM). He says that living in the United States has been very exciting for him. He is really happy that he has been able to meet people from many different cultures. Peter says that he has met people from all over the world, who are living in the Los Angeles, California area. He has made many new friends from India, Europe, Africa, South America, the United States, and other Countries. Peter is the son of Peter Mosua Lobe, II, and Susan Bosari Mulango. He is the first-born of four children, Minet Mosua, Elvis Mosua, and Bridget Nalenga, respectively.

Peter and my other African friends emphasized the importance of education, and learning about the people, and the various African Countries. They all enjoy the United States and meeting people from other countries who live here. My friends said they hope that African Americans will visit Africa, if possible.

Peter Lobe, from Cameroon

Kenya Ports Authority

Evelyn Knight & Long Beach Delegation visiting members of the Masai Tribe at the Masai Mara National Reserve in Kenya

Mombasa/Long Beach Sister Cities

I am very happy to be a part of the formation of a new Sister City for Long Beach, California. There were already seven Sister Cities in Long Beach, and the new one is Mombasa, Kenya. I feel that this connection to Africa will be great, and that it is long overdue. Our Long Beach Sister Cities are:

- **Bacalod, Philippines**
- **Izmir, Turkey**
- **Manta, Ecuador**
- **Mombasa, Kenya**
- **Phnom Penh, Cambodia**
- **Quingdao, China**
- **Sochi, Russia**
- **Yokkaichi, Japan**

A group of our Long Beach members visited the Mombasa Sister City at the end of 2008. I was unable to make the trip. One of my friends went with the group, and I asked her if she would take plenty of pictures, and give me a review of the trip. She took lots of pictures, and gave me the following information.

My friend Evelyn Knight, from Long Beach, California

The Long Beach Inaugural Delegation visited the City of Mombasa, Kenya from November 29, to December 11, 2008. The Sister Cities, Inc. of Long Beach approved the application of the eighth Sister City of Long Beach, in September, 2007. The City Council of Long Beach approved the Sister City relationship with Mombasa, Kenya, on (October 26, 2007), and this was the first visit to the Sister City. The group consisted of Founding Member and President of the Long Beach Mombasa Sister City, Bill Preston. Other Founding Members attending were, Evelyn Knight, Jesse Johnson (Founder and President of the 100 Black Men of Long Beach, Inc.), and Julie Hatter. Beverly Ragland, and Patricia Maye were also a part of the Long Beach Delegation visiting Mombasa.

The purpose of the visit to the Sister City was to implement the mission of the Sister City Association. The mission is to create rapport between the Cities by exchanging cultural, civic, educational, commercial, and industrial activities and relationships. The Cities of Long Beach and Mombasa have a great deal in common. They are both coastal cities with very successful ports, and both have very important roles in their areas. Both Cities have a population of about a half-million people. Each city also has an extremely diverse population, and they have a great attraction for tourists. Mombasa is on the Indian Ocean, in East Africa, and is popular for its white sandy beaches, and world class hotels and warm friendly people.

Our delegation provided about 50 books to the Kenya National Library. These books were donated by organizations, groups and individuals from Long Beach. The delegation participated in numerous events while they were in Mombasa. First, they attended a World Aids Prevention and Education Project. An Orphanage was also visited, and supported with gifts by the group. A workshop on Educational, Business Exchange Opportunities was presented. The Mombasa Long Beach Delegation was very engaged all during the visit. All members of the Long Beach Delegation were presented with gifts from the Sister City. The delegation had an opportunity to visit Kenya's game parks and reserves, which are the most outstanding and well known in the world.

The Public Officials we visited in Mombasa expressed an interest and commitment to visit their Sister City of Long Beach for cultural, civic, educational, and business exchange. They hope to visit in the near future.

On February 21, 2009, a delegation of six people from Mombasa, Kenya visited the Long Beach, California Library. The Main Library in Long Beach sponsored a 'Black History Month' program that day, and the delegation was included in the celebration. All members of the delegation were presented with gifts from the Main Library, the Mombasa Sister City, the African American Heritage Society of Long Beach, Congress Member Laura Richardson, The 100 Black Men of Long Beach, Inc., and the Port of Long Beach. Gifts from the Mombasa delegation were also presented. The delegation from Mombasa included:

Deputy Mayor, John McHaro
Town Clerk, Tubman Otieno
Public Relations Coordinator, Linet Mavu
Municipal Surveyor, Rose Manupe
Municipal Engineer, Albert Keno
Municipal Social Services Chairman, Mamun Abubakarali

Kenya National Library Service - camel picture, bus picture

Book deliveries are made to the various villages

Top: At restaurant in Kenya, L-R, Tour Guide, Bill Preston, Jesse B. Johnson, Jr., Patricia Maye, Tour Guide, Julie Hatter, Beverly Ragland, Evelyn Knight

Bottom: Books presented to Mombasa Library by Long Beach Delegation. L-R, Bill Preston (Founder & President Mombasa/Long Beach Sister Cities), Mombasa Librarian, Evelyn Knight (Founding Member MLBSC), Jesse B. Johnson, Jr., (Founding Member, MLBSC), Mombasa Librarian

Top - Likoni Aids Orphanage - Mombasa, Kenya

Bottom

Mombasa Kenya Delegation at Main Long Beach Public Library
Sister Cities of Long Beach Collection – Established in 2010

L-R, Linet Mavu, Tubmun Otieno, Albert Keno, Aaron L. Day, Mamun Abubakarali, John McHaro, Rose Manupe, Evelyn Knight, Bill Preston

Evelyn Knight at Masai Mara National Reserve in Kenya

Evelyn's Birthday party & Tribal Group pictures

**Mombasa, Kenya Delegation in Long Beach, California.
Top: L-R, Evelyn Knight, Herb Levi, Bill Preston, Linet Mavu, Tubmun Otieno, Mamun Abubakarali, Beverly Ragland, John McHaro, Rose Manupe, Albert Keno**

Bottom: L-R, Linet Mavu (Public Relations Coordinator), Rose Manupe (Municipal Surveyor), Albert Keno (Municipal Engineer), Tubmun Otieno (Town Clerk), John McHaro (Deputy Mayor), Mamun Abubakarali (Municipal Social Services Chairman), Congresswoman Laura Richardson, 37th District California

BOOKS PRESENTED TO MOMBASA. KENYA
by
LONG BEACH INAUGURAL DELEGATION - NOVEMBER 2008

- *100 African Americans Who Shaped American History*, Chrisanne Beckner
- *Cultures of America: African Americans*, Shelia Payton
- *America's Black Congressmen*, Maurine Christopher
- *Black Americans of Achievement: Duke Ellington*, Ron Frankl
- *Women of Color In The United States: A Guide to the Literature*, Bernice Redfern
- *Locating Free African American Ancestors: A Beginner's Guide*, Aaron L. Day
- *Great Negroes: Past and Present*, Russell L. Adams
- *Life and Times of Frederick Douglass*, Frederick Douglass
- *Coretta Scott King*, Jill C. Wheeler
- *Black Women In White America*, Gerda Lerner
- *The Autobiography of Malcolm X*, Alex Haley
- *African Americans*, Richard A. Long
- *From Slavery To Freedom: A History of Negro Americans*, John Hope Franklin
- *The Black American*, Leslie H. Fishel, Jr., /Benjamin Quarles
- *Afro-American History: Primary Sources*, Thomas R. Frazier
- *The Heritage of African Americans In Long Beach: Over 100 Years*, Aaron L. Day/Indira Tucker
- *The Measure of our Success: A Letter to My Children and Yours*, Marian Wright Edelman
- *The Heart of a Woman*, Maya Angelou
- *His Eye is on the Sparrow*, Ethel Waters
- *My American Journey*, Colin Powell with Joseph E. Persico
- *Stories From The Past*, Autrilla Watkins Scott
- *Bigmama Didn't Shop At Woolworths*, Sunny Nash
- *Constitution of United States of America & Constitution of State of California*, Darryl R. White
- *Art in America*, October 2007, and April 2008 Editions
- *Holy Bible*, The Gideons International (2 copies)
- *Untold Legacies: Pictorial History Black Long Beach 1900-2000 & Beyond*, NCNW/Sunny Nash
- *Sally Hemings*, Barbara Chase Riboud
- *Sunday Brunch*, Norma Jarrett
- *African-American Art: The Long Struggle*, Crystal A. Britton
- *Lessons in Living*, Susan L. Taylor
- *In My Father's House*, Ernest J. Gaines
- *The Underground Railroad*, R. Conrad Stein
- *Slavery In America*, Dorothy Schneider & Carl J. Schneider
- *Long Beach: A 21st Century City by the Sea*, Pioneer Publications, Inc.
- *Freedom & Sacrifice: Evelyn Knight & Cesar Chavez* Burnett Elementary School
- *History Lessons*, Aaron L. Day
- *Conflict in Black & White*, Dr. Charles Shockley

BOOKS, LIBRARY MATERIALS, GIFTS TO MOMBASA, DONATED BY:

- Long Beach Times Newspaper
- 100 Black Men of Long Beach
- KSUN Incorporated
- Marvin Colburt
- Julie Hatter-Jewels of the World Travel
- Sharon McLucas
- AY Roots
- Phyllis Venable
- Port of Long Beach
- Bonnie Lowenthal
- Village Treasures
- Autrilla Watkins Scott
- Long Beach Sister Cities Inc.
- Doris Topsy-Elvord
- The Breakers of Long Beach
- Ahmed Saafir
- Long Beach Public Library
- Long Beach Public Library Foundation

- Bill Preston
- NAACP, Long Beach Branch
- Evelyn Knight
- Press Telegram of Long Beach
- Indira Hale Tucker
- National Council for Negro Women
- Aaron L. Day
- Historical Society of Long Beach
- Jesse Johnson
- Long Beach Black Chamber of Commerce
- Andrew Mwangale
- Shore Books
- Marcus Tucker
- Maurice Oketch
- Sunny Nash
- African American Heritage Society
- Long Beach   Delta Sigma Theta Sorority Long Beach
- John Malveaux

Mombasa Delegation and Mayor Bob Foster, Long Beach, California

L-R, Bill Preston, Rose Manupe, Tubmun Otieno, Mayor Bob Foster (Long Beach), Vice Mayor John McHaro (Mombasa), Albert Keno, Linet Mavu, Mamun Abubakarali, Beverly Ragland, Laverne Duncan, Melissa Morgan

Again, I feel that this connection to Africa will be great, and that it is long overdue. Our City of Long Beach is very happy to have this connection to Africa. We had another delegation from Long Beach to Mombasa, in November, 2009.

Mombasa Delegation visits Long Beach City Hall

L-R, Tubmun Otieno, John McHaro, Albert Keno, Mamun Abubakarali, Linet Mavu, Beverly Ragland, Rose Manupe, Bill Preston

PART III

TECHNIQUES AND TIPS:

AFRICAN AMERICAN

RESEARCH RESOURCES

RESEARCHING APPRENTICE BONDS

```
Apprentice                                Bound to           Date
Day, George -4 (son of Nan Day)           Samuel Winstead    20 Jun.1780
Day, George (of colour)                   J. S. Hutchinson   12 Jul.1820
Day, Jessie (daughter of Rachel Day)
        6 months                          Drury Allen        20 Mar.1780
Day, Lucy -1 ( child of Anne Day)            "                   "
Day, Thom (son of Rachel) -3              Samuel Winstead    20 Jun.1780
```

RESEARCHING MARRIAGE BONDS

PERSON COUNTY MARRIAGE BONDS

Page 47

GROOM	BRIDE	DATE OF BOND	BONDSMAN & WITNESS
Day, John	Jane Sweaney	19 Jan. 1824	Robt. D. Bumpass, (Jr.)
Day, John	Mary Cozart	7 July 1830	Alexander Lunsford
Day, John x	Jane Mitchell	9 May 1839	David Ramsey (w) Ch. Mason
Day, John	Susan A. Ellixson	12 Feb. 1867	Thomas C. Day (w) Saml. Y. Brown,Clk. m. 13 Feb. 1867 by D. R. Moore
Day, Washington (Col.)	Milly Day (Col.) (dau. of Scott & Letta)	25 Oct. 1867	(w) Sam'l Y. Brown,Clk. m. 28 Oct. 1867 by B. A. Thaxton, J. P.

CHAPTER 10

RESEARCH STEPS

LEARNING MORE ABOUT YOUR ANCESTORS

Review of 15 research steps

1. Start with yourself

2. Interview and get information from relatives

3. Pedigree/Ancestor chart

4. Family Group sheet

5. Research log/calendar

6. Vital records

7. Check counties and states where ancestors lived

8. Federal Records – genealogical, census, military, ship passenger arrival and departure lists, land records, etc.

9. Libraries

10. Genealogical societies

11. Internet

12. Books/magazines

13. DNA research

14. Organization of files

15. Sharing with others

☐ **Start with yourself** – Collect what you already know about your parents, grandparents, great-grandparents and other ancestors. Start with family documents, letters, bibles, obituaries, marriage records, photographs, deeds, wills, etc.

☐ **Interviewing and getting information from relatives** – Learn more about your ancestors by collecting information, documents, mementos and other items from relatives. Also, interview older living family members for information.

☐ **Pedigree/ancestor chart** – Record information located about parents, grandparents, and great-grandparents on these charts. These charts trace the family lines back for both parents.

☐ **Family group sheet** – Do a family group sheet for each marriage. This information is very helpful when researching ancestors on the pedigree/ancestor charts.

☐ **Research log/calendar** – You will be able to keep track of documents and records that you have researched by keeping careful records.

☐ **Vital records** – Birth, death, and marriage records are good sources of information about ancestors.

☐ **Checking counties and states where ancestors lived** – Information may be obtained about ancestors by checking records and documents in the local counties, surrounding counties, and states where they lived.

☐ **Federal records** – The National Archives and Records Administration contains information about individuals whose names appear in census records, military service and pension files, ship passenger arrival and departure lists, as well as other records.

☐ **Libraries** – Many libraries carry genealogical collections and books on doing family history research. It is very important to check the collections at different libraries, as they all carry a wide range of genealogical materials. Also, check the LDS Family History Libraries.

☐ **Genealogical societies** – Anyone interested in learning about the field of genealogy and the basic elements and terminology of genealogical research is encouraged to join a local society.

☐ **Internet** – Access to genealogical information on the Internet has greatly improved over the past decade. Many Web sites offer tips on researching family histories and have compiled lists, which have been submitted by family historians.

☐ **Books/magazines** – There is an enormous amount of information to be found in genealogical books and magazines. Subscribers submit stories about their ancestors, there are interesting genealogical articles, and how-to-tips, and there is always up-to-date information about the Internet.

☐ **DNA research** – Using DNA testing to discover ancestral backgrounds is the latest technology for your genealogy research. Today, researchers may find out if two people are related, discover if two people descended from a common ancestor, obtain clues about ethnic origins, etc.

☐ **Organization of files** – Researchers soon learn that it is very important to organize their files and data so they can be easily retrieved. Organization of files is also very important when you are working on various family groups.

☐ **Sharing with others** – Remember to share your information with other family members and friends. As you accumulate information about your ancestors, be sure to pass it on for others to enjoy. Your descendants will appreciate the fact that you cared enough to help pass on the family legacy. You owe it to your ancestors, and to your descendants.

RECORDS FOR SLAVE RESEARCH - (BEFORE 1870)

CHALLENGES: Few written records, name changes, nicknames, and migration

* Checking records to locate slave owner

* Federal Census Schedules, 1850 and 1860 (Slave Schedules)

* The 1850 and 1860 Federal Census (Mortality Schedules)

* Tax Records & Will Records

* Land and Property Records

* Probate, Estate, and Chancery Court Records

* Plantation Records

* Antebellum Southern Plantations from the Revolution through the Civil War

* Bills of Sale

RECORDS FOR FREE ANCESTOR RESEARCH - (BEFORE 1870)

* Census Schedules

* Higher Court Petitions

* Marriage Bonds

* Manumissions

* Apprentice/Indenture Bonds

* Free Register Records

* Land Deeds

* Tax Records

* Court Records (Status of Mother)

* Military Records

SEARCHING RECORDS - (AFTER 1870)

*   Census Schedules

*   United States Military Records

*   Homestead Records

*   Cemetery Records

*   City Directories

*   Church Records

*   Court Records

*   Land Records

*   Probate Records

*   School Records

 *   Vital Records - (government records of births, marriages, deaths, and Social Security Death Index)

SEARCHING TRANSITION RECORDS FROM
SLAVERY TO FREEDOM

*   Records of the Underground Railroad

*   Records that help you make the Slave Connection

*   Military Records

*   Freedman's Savings and Trust Records

*   Bureau of Refugees, Freedmen, and Abandoned Lands (Freedmen's Bureau)

*   Southern Claims Commission Records

*   American Missionary Association Records

*   Check ways of identifying the slave owner

TYPES OF DOCUMENTS USED TO LOCATE THE DAY ANCESTORS

* **School Records**

* **Church Records**

* **Family Reunions/Weddings/Other Events**

* **Photos**

* **Medical Records**

* **Birth Certificates**

* **Marriage Records**

* **Death Records**

* **Tax Records**

* **Social Security Records**

* **Wills**

* **Military Records**

INCLUDING THE FOLLOWING:

ORAL HISTORY

BOOKS/STATE PAPERS

U.S. CENSUS SCHEDULES

MARRIAGE BONDS

APPRENTICE/INDENTURE BONDS

EARLY TAX RECORDS

LAND ENTRY DEEDS

PARISH RECORDS

COURT RECORDS

STATE/COUNTY RECORDS

EARLY BIRTH RECORDS

TRANSPORTATION RIGHTS

FREE REGISTER RECORDS

Day Households in the United States, 1790-1830

Checking your free ancestors through the U.S. Census Schedules may be very helpful. If your ancestors were living in the United States in 1860, you may be able to locate them on the Census Schedules, if they were free, as I did with my Day ancestors. The 1870 Census included slaves and free persons of color, as the Civil War had just ended in 1865.

During my research, I found that many first names were used from generation to generation, and it can become very confusing. For instance, I found many family members named: John, James, George, Thomas, Rachel, Nancy, Milly, Winnifred, etc. In order to distinguish the different generations, I have given many of these family members an initial. I usually use the first letter of their mother's or father's first name. For instance, the son of John who is also named John will be (John J. Day), or Nancy who is named after a mother or grandmother will be (Nancy N. Day). Some names have different spellings, which also helps to distinguish that person. I have found Milly spelled (Millie), Winnifred spelled (Winifred), and Rachel spelled (Rachell), so, I will use one spelling for one generation, and the other spelling for another generation.

After tracing my great-great-grandfather Scott, back to the 1830 Census, I continued checking back to 1790. In the 1830 Census Schedule, I located 1,437 Day households in the United States. When I got back to the 1790 Census, I discovered that there were only 322 Day households in the United States during that period, and they were living in 12 States. Some of these Days are listed as **W-** for White, **B -** for Black, **C-** for Colored, and **M-** for Mulatto (mixed parentage).

I checked each State and extracted a total number of 322 Day Households living in 12 States in 1790. Only the heads of the households are listed by name. You can begin this process by:

- Checking the 1870 Census and working back to 1790 (first census)
- Checking each State for your surname
- Checking the Counties where you locate your ancestors

If you are tracing Day ancestors, the following may be helpful to your research, and may also be used as a guideline for those who are tracing different family names.

'Day Households in the United States, 1790-1830

YEAR	DAY HOUSEHOLDS	STATES
1790	322	12
1800	429	13
1810	661	15
1820	1,042	23
1830	1,537	26

The 1790 U.S. Census Schedules for Delaware, Georgia, Kentucky, New Jersey, Tennessee, and Virginia were apparently destroyed during the British attack on Washington during the War of 1812. The 1790 Schedules for Virginia were reconstructed from State Enumerations, and appear on Microfilm Publication T498.

Day Households (Families) in the United States - 1790

1. CT - 40	7. NC - 19
2. ME - 2	8. PA - 31
3. MA - 99	9. RI - 1
4. MD - 21	10. SC - 21
5. NH - 24	11. VT - 25
6. NY - 30	12. VA - 9

322

By 1800, the number of Day Households in North Carolina where I located so many of my ancestors, had increased to a total of thirty. They were now living in 10 different counties. The total Day population had increased to 429, in fourteen States.

Day Households (Families) in the United States – 1800

1. CT	-	42	8. NY	-	61
2. DE	-	2	9. NC	-	30
3. DC	-	1	10. PA	-	34
4. ME	-	49	11. RI	-	3
5. MD	-	1	12. SC	-	23
6. MA	-	116	13. VT	-	29
7. NH	-	29	14. VA	-	9
					————
					429

By 1810, the number of 'Day Households in North Carolina had increased to a total of forty one. They had now branched to fourteen different counties. The total Day population had increased to 661 fifteen States.

Day Households (Families) in the United States - 1810

1. CT - 45		8. NH - 28
1. CT - 43		9. NY - 137
2. DE - 9		10. NC - 41
3. KY - 35		11. PA - 45
4. LA - 2		12. RI - 3
5. ME - 80		13. SC - 26
6. MD - 23		14. VT - 48
7. MA - 132		15. VA - 9

661

By 1820, the number of 'Day Households in North Carolina had increased to a total of forty eight. They were still living in fourteen different counties. The total Day population had increased to 1,042, in twenty three States

Day Households (Families) in the United States - 1820

1. CT - 45	12. MI - 2
2. DC - 4	13. MS - 8
3. DE - 9	14. NH - 29
4. GA - 14	15. NY - 179
5. IL - 7	16. NC - 48
6. IN - 13	17. OH - 100
7. KY - 51	18. PA - 72
8. LA - 10	19. RI - 4
9. ME - 108	20. SC - 26
10. MD - 26	21. TN - 15
11. MA - 133	22. VT - 52
	23. VA - 87
	_______
	1,042

By 1830, the number of 'Day Households in North Carolina had increased to a total of sixty. They were now living in fifteen different counties. The total Day population had increased to 1,427, in twenty six States.

Day Households (Families) in the United States – 1830

1.	AL	- 22	14.	MO	- 6
2.	CT	- 55	15.	MS	- 16
3.	DC	- 11	16.	NC	- 58
4.	DE	- 11	17.	NH	- 85
5.	GA	- 24	18.	NJ	- 80
6.	IL	- 5	19.	NY	- 312
7.	IN	- 30	20.	OH	- 112
8.	KY	- 43	21.	TN	- 15
9.	LA	- 8	22.	RI	- 5
10.	MA	- 147	23.	SC	- 34
11.	ME	- 145	24.	TN	- 16
12.	MI	- 6	25.	VT	- 50
13.	MD	- 35	26.	VA	- 106

1,537

LIBRARIES/GENEALOGY SOCIETIES/MUSEUMS

Some libraries have excellent collections of genealogical and African American materials. Libraries are great places to research family histories. The following updated list includes many of the well-known libraries that have excellent genealogical materials and other resources, but is not a complete list. Genealogy Societies and Museums of interest have also been added.

Four grandchildren of Aaron L. Day, at Wilberforce, Ohio Museum and Cultural Center, L-R, Ephrom Moody, Crystal Moody, Ray Moody, Jr., (front) - Arrionne Moody)

Aaron's mother Lula Banks-Day, graduated from Wilberforce University in 1931

Alabama

Alabama A M University
P.O. Box 489
Normal, AL 35762
PHONE: (256) 372-5846
FAX: (256) 372-5338

Alabama State University
Ollie Brown A. A. Collection
Montgomery, AL 36104
PHONE: (334) 229-4100
E-mail: webmaster@asunet.alasu.edu

Birmingham Public Library
2100 Park Place
Birmingham, AL 35203-2974
PHONE: (205) 226-3610
http://www.bplonline.org

Hollis Burke Frissell Library
Tuskegee, AL 36088
PHONE: 1-800-622-6531
webmaster@tusk.edu

Mobile Public Library
Local History & Genealogy Division
700 Government Street
Mobile, AL 36602-1403
PHONE: (251) 208-5865
FAX: (251) 5865
E-mail: mobilepl@acan.net

Arizona

Afro-American Historical &
Genealogical Society of Tucson
7739 E. Broadway, Suite #195
Tucson, AZ 85710

Arizona State Archives
1700 W. Washington St.
Phoenix, AZ 85007
PHONE: (602) 542-4159
FAX: (602) 542-4402
E-mail: archive@dlapr.lib.az.us

Arkansas

Afro-American Historical and
Genealogical Society, Inc.
P.O. Box 4294
Little Rock, AR 72214
FAX: (501) 225-2029
E-mail: ttenplewis@aol.com

Southwest Arkansas Regional Archives
201 Highway 195 South, P.O. Box 134
Washington, AR 71862
http://www.southwestarchives.com

California

Mayme A. Clayton Library & Museum
4130 Overland Avenue
Culver City, CA 90230
PHONE: (310) 202-1647
www.claytonmuseum.org

A C Bilbrew Library
150 E. El Segundo Blvd.
Los Angeles, CA 90061
PHONE: (310) 538-3350
http://www.colapublib.org/libs/bilbrew/brc.ht

African American Genealogical Society of
Northern California
Oakland Public Library/Diamond Br.
3565 Fruitvale Ave.
Oakland, CA 94602-2398
http://www.aagsnc.org/organization.htm

African American Genealogy Society of
Sacramento, California
P. O. Box 277681
Sacramento, CA 95827-7681
http://www.aagssc.org

African American Heritage Society
Of Long Beach
P. O. Box 5903
Long Beach, CA 90805
http://www.lbpl.org/aahs.html

California African American
Genealogical Society
P.O. Box 8442
Los Angeles, CA 90008-0442
E-mail: zmerald@netzero.net

Huntington Beach City Library
711 Talbert Ave.
Huntington Beach, CA 92648
PHONE: (714) 842-4481
FAX: (714) 375-5180

Long Beach Public Library
101 Pacific Ave.
Long Beach, CA 90822
PHONE: 570-7500
http://www.lbpl.org

Los Angeles Public Library
630 W. Fifth Street
Los Angeles, CA 90071
PHONE: (213) 228-7000
FAX: (213) 228-7069
http://www.lapl.org

Los Angeles Temple
Family History Center
10741 Santa Monica Blvd.
West Los Angeles, CA 90024
PHONE: (310) 474-9990
http://www.familysearch.org

Oakland Public Library
Martin Luther King Jr., Br.
6833 International Blvd.
Oakland, CA 94621
PHONE: (510) 615-5728

San Diego African American
Genealogy Research Group
P.O. Box 740240
San Diego, CA 92174-0240
PHONE: (619) 262-5810

Southern California Genealogical
Society & Library
417 Irving Drive
Burbank, CA 91504-2408
PHONE: (818) 843-7247
E-mail: scgs@scgsgenealogy.com

Colorado

Black Genealogical
Research Group
4605 E. Kentucky Avenue, 5F
Denver, CO 80222
http://www.pbs.org/kbyu/ancestors/resourcegui

Denver Public Library
10 W. Fourteenth Ave. Pkwy.
Denver, CO 80204
PHONE: (720) 865-1111
http://www.denver.lib.co.us

Connecticut

Connecticut Afro-American
Historical Society
444 Orchard Street
New Haven, CT 06511

Connecticut Historical Society
One Elizabeth Street
Hartford, CT 06105
PHONE: (860) 236-5621
E-mail: ask_us@chs.org

Connecticut State Library
231 Capitol Avenue
Hartford, CT 06115
PHONE: (860) 757-6500
http://www.cslib.org

Prudence Crandall
Museum Library
P.O. Box 58
Canterbury, CT 06331
PHONE: (860) 546-9916

District of Columbia

Afro-American Historical and
Genealogical Society
P.O. Box 73086
Washington, DC 20056-3086
www.rootsweb.com/~mdaahgs/

Afro-American Historical and
Genealogical Society
National Capitol Area
P.O. Box 34683
Washington, DC 20043

Library of Congress
Local History and Genealogy Division
101 Independence Ave, S E
Washington, DC 20540
PHONE: (202) 707-5000

Moorland-Spingarn Research Center
Howard University
500 Howard Place, NW
Washington, DC 20059
PHONE: (202) 806-7240
FAX: (202) 806-6405

Martin Luther King Jr.
Memorial Library
901 G Street, NW
Washington, DC 20001
PHONE: (202) 727-0321
http://dclibrary.org

National Society Daughters of the
American Revolution Library
1776 D Street NW
Washington, DC 20006-5303

Florida

Afro-American Historical and
Genealogical Society of
Central Florida
P.O. Box 5742
Deltona, FL 32728

Bethune-Cookman College
640 Dr. Mary McLeod Bethune Blvd.
Daytona Beach, FL 32114-3099
PHONE: (386) 481-2000

Florida Agricultural & Mechanical Univ.
University Library
Tallahassee, FL 32307
http://www.famu.edu

Georgia

Atlanta-Fulton Public Library
Special Collections Department
One Margaret Mitchell Square
Atlanta, GA 30303
PHONE: (404) 730-1700
http://www.af.public.lib.ga.us

Genealogical Center Library
P.O. Box 71343
Marietta, GA 30007-1343

Robert W. Woodruff Library
Atlantic University Center
Black Culture Collection
111 James P. Brawley Drive, SW
Atlanta, GA 30314
PHONE: (404) 522-8980
http://www.auctr.edu

Sara Hightower Regional Library
Special Collections
205 Riverside Parkway
Rome, GA 30161
PHONE: (706) 236-4600
http://www.floyd.public.lib.ga.us

Illinois

African-American Genealogical Society of
Central Illinois
314 North Main
Decatur, IL 62523
PHONE: (217) 429-7458

Afro-American Genealogical and
Historical Society of Chicago
P.O. Box 377651
Chicago, IL 60637

DuSable Museum of African
American History Library
740 East 56th Place
Chicago, IL 60637-1495

International Society of Sons and
Daughters of Slave Ancestry
Box 436937
Chicago, IL 60643
www.rootsweb.com/~ilissdsa/

Newberry Library
60 West Walton St.
Chicago, IL 60610-7324
PHONE: (312) 943-9090
http://www.newberry.org/nl/genealogy/af-amer-

Woodson Regional Library of the
Chicago Public Library
9525 South Halsted Street
Chicago, IL 60628
PHONE: (312) 747-6900
http://www.uic.edu/orgs/cmhec/collectiondirec

Indiana

Allen County Public Library
900 Webster Street, P O Box 2270
Fort Wayne, IN 46802-2270
PHONE: (260) 421-1200
http://www.acpl.lib.in.us

Indiana African American Historical
and Genealogical Society
502 Clover Terrace
Bloomington, IN 47404-1809

Indiana Historical Bureau
140 North Senate Ave.
Indianapolis, IN 46204-2296
PHONE: (317) 232-2535
FAX: (317) 232-3728
E-mail: ihb@statelib.lib.in.us

Neal-Marshall Black Culture Center
Blog. 172, Rm. 113
275 N. Jordan Ave.
Bloomington, IN 47405
E-mail: aaclib@indiana.edu

Kansas

Leavenworth Afro-American
Historical Society
P.O. Box 3151
Fort Leavenworth, KS 66027
PHONE: (913) 651-4584

Louisiana

Archives of the Parish of
East Baton Rouge Genealogy Section
Public Service Department
222 St. Louis Street
Baton Rouge, LA 70802

New Orleans Public Library
Louisiana Division
219 Loyola Avenue
New Orleans, LA 70112-2044
PHONE: 529-7323
http://nutrias.org/~nopl/guides/genguide/ggco

Tulane University-Tilton Hall
6823 St. Charles Ave.
New Orleans, LA 70118
PHONE: (504) 865-5535
FAX: (504) 865-5580
E-mail: arc@tcs.tulane.edu

Maryland

Afro-American Historical and
Genealogy Society of Baltimore
P.O. Box 66265
Baltimore, MD 21218

Afro-American Historical and Genealogical
Soc. of Central Maryland
P.O. Box 2774
Columbia, MD 21045

Maryland Museum of African-
American History & Culture
100 Community Place
Crownsville, MD 21032
PHONE: (410) 514-7654
E-mail: smithn@dhcd.state.md.us

Massachusetts
American Antiquarian Society
185 Salisbury Street
Worcester, MA 01609-1634
PHONE: (508) 755-5221
http://www.americanantiquarian.org

New England Historic
Genealogical Society
101 Newbury Street
Boston, MA 02116-3007

Mildred F. Sawyer Library
Suffolk University
8 Ashburton Place
Boston, MA 02108
http://www.suffolk.edu/admin/sawlib/sawyer.ht

Michigan

Center for Afro-American and
African Studies
550 East University Street
Ann Arbor, MI 4109
PHONE: (734) 764-5518
http://www.umich.edu/niinet/caas/fellowships96

Central Michigan University
Clarke Historical Library
Mt. Pleasant, MI 48859
PHONE: (989) 774-4000
http://www.cmich.edu

Charles H. Wright
Museum of African American History
315 E. Warren Ave.
Detroit, MI 48201-1443
PHONE: (313) 494-5800
FAX: (313) 494 5855

Detroit Public Library
5201 Woodward Ave.
Detroit, MI 48202
PHONE: (313) 833-1480
http://www.detroit.lib.mi.us/burton

Missouri

Black Archives of Mid-America
2033 Vine
Kansas City, MO 64108
PHONE: (816) 483-1300
http://www.blackarchives.org

St. Louis University of Missouri
Thomas Jefferson Library
8001 Natural Bridge Road
St. Louis, MO 63121
PHONE: (314) 516-5000
http://www.umsl.edu

Mid-Continent Public Library
317 W. 24 Highway
Independence, MO 64050
http://www.mcpl.lib.mo.us

University of Missouri-Columbia
23 Ellis Library
Columbia, MO 65211
PHONE: (573) 882-2121
E-mail: webeditor@missouri.edu

Nebraska

Great Plains
Black Museum Library
2213 Lake Street
Omaha, NE 68110
PHONE: (402) 345-2212
http://www.omaha.org/oma/black.htm

Nebraska State Historical Society
1500 R Street
Lincoln, NE 68501
http://www.nebraskahistory.org

New Hampshire

New Hampshire Historical Society
30 Park Street, P.O... Box 478
Concord, NH 03301
PHONE: (603) 228-6688
http://www.nhhistory.org

New Hampshire State Library
20 Park Street
Concord, NH 03301
PHONE: (603) 271-2144
http://www.state.nh.us/nhsl

New Jersey

Afro-American Historical and
Genealogical Society of New Jersey
1841 Kennedy Boulevard
Jersey City, NJ 07305

Danforth Memorial Library
250 Broadway
Paterson, NJ 07501
PHONE: (973) 357-3000
FAX: (973) 881-8338
http://www.nationalregisterofhistoricplaces.c

Afro-American Historical
Society Museum
1841 Kennedy Boulevard, 2nd Fl.
Jersey City, NJ 07305
PHONE: (201) 547-5262
FAX: (201) 547-5392
E-mail: info@cityofjerseycity.org

New York

African-American Cultural Center
350 Masten Avenue
Buffalo, NY 14209
PHONE: (716) 884-2013
FAX: (716) 885-2590
E-mail: aacc@pcom.net

Afro-American Historical and
Genealogical Society
P.O. Box 022340
Brooklyn, NY 11202-0049
PHONE: (212) 330-7882
www.rootsweb.com/~nynassa2/afroamerica

New York Public Library
Humanities & Social Sciences Library
Fifth Avenue and 42nd Street
New York, NY 10018
PHONE: (212) 930-0828
www.nypl.org/research/chss/lhg/genea.html

New York Public Library
Schomberg Center
515 Malcolm X Boulevard
New York, NY 10037-1801
PHONE: (212) 491-2200
http://www.nypl.org/research/sc/sc.html

Staten Island Institute of Arts and
Sciences Archives and Library
75 Stuyvesant Place
Staten Island, NY 10301
PHONE: (718) 727-1135

University of Rochester
Government Documents
Rush Rhees Library
Rochester, NY 14627
PHONE: (585) 275-2121

North Carolina

Afro-American Genealogical
Society, North Carolina
P.O. Box 26785
Raleigh, NC 27611-6785

Mattye Reed African Heritage Center
A & T State University, Dudley Bldg.
Dudley Street
Greensboro, NC 27411
PHONE: (336) 334-3209
FAX: (336) 334-4378

North Carolina African-American
Heritage Society
P.O. Box 26334
Raleigh, NC 27611

Olivia Raney
Local History Library
4016 Carya Drive
Raleigh, NC 27610
PHONE: (919) 250-1196

Ohio

Ohio Historical Society
1982 Velma Ave.
Columbus, OH 43211
PHONE: (614) 297-2340
http://www.ohiohistory.org

African-American Genealogical
Society of Cleveland
P.O. Box 200382
Cleveland, OH 44120-9998

Cleveland Public Library
History and Geography Department
325 Superior Avenue
Cleveland, OH 44114-1271
PHONE: (216) 623-2800

National Afro-American
Museum and Cultural Center
Box 578
Wilberforce, OH 45384-1001
1-800-752-2603
http://www.ohiohistory.org/places/afroam/

Ohio State University
Black Studies Library
1858 Neil Avenue Mall
Cleveland, OH 43210-1286
http://www.lib.ohio-state.edu

Public Library of Cincinnati
& Hamilton County
800 Vine Street
Cincinnati, OH 45202-2071
PHONE: (513) 369-6900
http://www.cincinnatilibrary.org

Wilberforce University
Special Collections
P.O. Box 1001
1055 N. Bickett Road
Wilberforce, OH 45384-1001
PHONE: (937) 376-2911

Oklahoma

Langston University
Black Heritage Center
P.O. Box 730
Langston, OK 73050
PHONE: (405) 466-3836
http://www2.luresext.edu

Pennsylvania

African American
Genealogy Group
P.O. Box 27356
Philadelphia, PA 19118
(215) 572-6063
info@aagg.org

Blockson Afro-American Collection
Temple University
Sullivan Hall, First Floor
12th Street and Berks Mall
Philadelphia, PA 19122
PHONE: (215) 204-6632
http://www.temple.edu/blockson

Historical Society of
Pennsylvania Library
1300 Locust Street
Philadelphia, PA 19107
PHONE: (215) 732-6200

Lincoln University
Special Collections
P.O. Box 179
Lincoln University, PA 19352-0999
PHONE: (610) 932-8300

Temple University
Charles L. Blockson Collection
1801 North Broad St.
Philadelphia, PA 19122
PHONE: (215) 204-7000

South Carolina

Avery Research Center
College of Charleston
Charleston, SC 29424
PHONE: (803) 727-2009
FAX: (803) 727-2017

South Carolina Historical Society
100 Meeting Street, Fireproof Bldg.
Charleston, SC 29401
FAX: (843) 723-8584
http://www.schistory.org

York W. Bailey Museum
P.O. Box 126
Saint Helena Island, SC 29920
PHONE: (843) 838-2432

Tennessee

African American Cultural Alliance
1819 Charlotte Avenue
Nashville, TN 37203
PHONE: (615) 299-0412
http://www.africanamericanculturalalliance.co

Center for Southern Folklore Archives
130 Beale Street
Memphis, TN 38101
PHONE: (901) 438-5855

Fisk University
Special Collections Department
Seventeenth at Jackson Street
Nashville, TN 37203
PHONE: (615) 329-8500

Texas

African American Genealogical and
Historical Society
P.O. Box 200784
San Antonio, TX 78220
http://www.co.bexar.tx.us/general/genealogy/g

Afro-American Historical and
Genealogical Society of Texas
P.O. Box 670045
Houston, TX 77267-0034

Black Genealogy
P.O. Box 6825
Lubbock, TX 79493-6825

Clayton Library Center
For Genealogical Research
5300 Caroline Street
Houston, TX 77004-6896
PHONE: (832) 393-2600
http://www.houstonlibrary.org/clayton

Dallas Public library
1515 Young Street
Dallas, TX 75201
PHONE: (214) 670-1424
http://www.dallaslibrary.org/CHS/cgc.htm

Tarrant County Court House
Black History
100 E. Weatherford St.
Ft. Worth, TX 76196
PHONE: (817) 884-1111

Utah

Family History Library
The Church of Jesus Christ of
Latter-day Saints
35 North West Temple
Salt Lake City, UT 84150-3400
http://www.lds.org

Registry of Black American Ancestry
P.O. Box 417
Salt Lake City, UT 84110

Virginia

Afro-American Historical and
Genealogical Society
P.O. Box 2448
Newport News, VA 23609-2448

Alexandria Black history
Research Center
638 North Alfred Street
Alexandria, VA 22314
PHONE: (703) 838-4829

http://www.cybvis.com/alex/abhrc.htm
Hampton University
William R. and
Norma B. Harvey Library
130 East Tyler Street
Hampton, VA 23668
PHONE: (757) 727-5000

The Library of Virginia
800 East Broad Street
Richmond, VA 23219-8000
PHONE: (804) 692-3500
http://www.lva.lib.va.us

Virginia Union University
Special Collections
1500 North Lombardy Street
Richmond, VA 23220
PHONE: (804) 257-5600
http://www.vuu.edu

Washington

Black Heritage Society
Of Washington State
P.O. Box 22961
Seattle, WA 98122-0961
PHONE: (2060 324-1126

Wisconsin

State Historical Society of
Wisconsin Library
816 State Street
Madison, WI 53706-1482

University of Wisconsin-Milwaukee
Department of History
P.O. Box 413 Holton Hall, 342
Milwaukee, WI 53201
PHONE: (414) 229-4361
http://www.uwm.edu/library/arch

INTERNET SITES

AAHGS (Nat'l   Afri-Am Gen Soc)
http://www.rootsweb.com/~mdaahgs

Afri-Am Genealogy Society of No Calif.
http://www.aagsnc.org

African/Native-American Connections
http://members.aol.com/angelaw859/index.html

Afrigeneas
http://www.afrigeneas.com

Black Cemeteries Online
http://www.prairiebluff.com/aacemetery

Black Civil War Soldiers (USCTs)
http://www.itd.nps.gov/cwss

Black Revolutionary War Info
http://americanrevolution.org/blk.html

Caribbean Resource List
http://www.candoo.com/genresources/index.html

Christine's list of Genealogy links
http://www.ccharity.com

Free Blacks of NC, VA, MD, DE, & PA
(Heinegg)
http://www.freeafricanamericans.com

Freedmen's Bureau Records
http://www.freedmensbureau.com

LDS
http://www.familysearch.org

Lest We Forget (Military, etc)
http://www.coax.net/people/LWF

Middle Passages Slave Ship Database
http://middlepassages.com

National Genealogical Society
http://www.ngsgenealogy.org

Plantation Slave Voices
http://scriptorium.lib.duke.edu/slavery/plantation.html

The Schomburg Center, NY
http://www.nypl.org/research/sc/sc.html

Southern   California Genealogical Society
http://www.scgsgenealogy.com

Underground Railroad Data
http://www.ugrr.org//research.html

US GenWeb – Archives
http://www.rootsweb.com/~usgenweb

Very helpful - CDs, Data, Books,

http://www.familytreemaker.com
http://www.ancestry.com
http://www.hearthstonebooks.com

TEACHING OUR YOUNGER GENERATIONS

Naomi Rainey, NAACP Long Beach Branch, President

I have especially enjoyed sharing research tips about genealogy/family history research to others through lectures, workshops, and field trips. Among these special experiences was my presentation to students from the **Jefferson Leadership Academies in Long Beach, California**. I represented the **NAACP Long Beach Branch**. In 2006, at the request of NAACP President Naomi Rainey, I gave a genealogy/family history research lecture to the class. The purpose of the class was to motivate the students into researching, documenting, and preserving their family histories. Their teacher, Mr. David Michaels, was very happy to know that his students would learn about the process of discovering and preserving their heritage, **(See Appendix III)**.

The project began when the Long Beach Branch NAACP adopted Jefferson Middle School through its Stay in School Program. The program was a recipient of grants from the **Arts Council for Long Beach, and Vons Foundation**. With their generous funds, the Long Beach Branch NAACP was proud to share with the community this product. As a result, this project was funded. The genealogical research used to collect information, and the creative writing used to bring this information to life, were performed by students participating in the Stay in School Program.

The Jefferson Leadership Academies School Principal, Helen Compton-Harris, was extremely grateful to the Long Beach Branch of the NAACP for their support of the school. Teacher David Michaels stated that "this opportunity gave the students a valuable lesson in family tradition, and it taught them how they could research their families' history.   Many students learned their real family history, and recognized that they may be the designated family member to keep all the family tree records for the future."

My goal was to inspire the students so that they might want to talk to older family members and learn about their heritage. Some of them may very well become the family historian. I outlined a long-range research plan for the students. This guideline is something that they will be able to use their entire lives, because family history research and preservation is a continuing process.

**   *   I told the students to begin by collecting photos, school records, family papers, birth records, obituaries, marriage records, and other documents. I also suggested that they interview their older relatives.**

***   With the assistance of their parents and other relatives, I suggested that they begin researching at libraries, and the counties where their ancestors had lived. They will be able to search through census schedules at the National Archives, the Family History Centers, and other Genealogical Institutions for information about the history of their ancestors. The internet is also available for research on family histories.**

***   To preserve this information for other family members and their descendants, I suggested that they begin putting this data into acid-free sheet protectors and binders.**

***   I also showed them how to begin filling out pedigree/ancestor charts, family group sheets, and to start writing about their personal experiences.**

I gave the students a 15-step plan that they could use through the coming years to learn more about their ancestors. Fifteen of the students were inspired to write about their families. Some of them filled out pedigree/ancestor charts and family group sheets. These students are very fortunate to be learning so much about their family heritage at such early ages. The students also submitted essays along with the charts and sheets. It was obvious from the work presented, that the students really do love and enjoy their families. They were inspired and wanted to learn more about the history of their ancestors.

The essays were warm and very touching. The love and pride seemed to flow through the pages. Not only are the students learning to research into the past, they are also documenting the present history by telling stories about themselves, and their family members. All of the students did excellent work on their essays and family history research. Remember that you are the link between the past and the future. **Why not consider becoming the family historian for your family - and pass the stories down to your descendants?**

Student Project-Family History – Book
www.lbcanaacp.org

100 Black Men of Long Beach, Inc.

Jesse B. Johnson, Jr., - 100 Black Men of Long Beach, Inc., Founder/President

A group of concerned African American men began to meet in New York in 1963 to explore ways of improving conditions in their community. These men were business and industry leaders such as Mr. David Dinkins, Mr. Robert Mangum, Dr. William <u>Hayling</u>, Mr. Nathaniel <u>Goldston</u> III, Mr. Livingston Wingate, Mr. Andrew Hatcher, and Mr. Jackie Robinson. The group eventually started the "100 Black Men of America, Inc." Today, the 100 Black Men of America is an international organization with 114 chapters located in the United States, England and the Caribbean.

In the year 2008, thanks to the hard work and dedication of 32 Charter Members, the 100 Black Men of Long Beach, Inc., was formed, **(See Appendix III).** This chapter is dedicated to improving the quality of life of those that live, work, and play in the greater Long Beach community. There is an emphasis on the African American family, with emphasis on **African American youth**. The "Four for the Future" (Economic Development, Education, Mentoring, Health and Wellness), as developed by the national body, are programs that have been implemented by the chapter to tackle the many challenges of the community.

One of the many projects the Long Beach group will be working on during the next few years is to help the young students learn more about their family histories. As chapter Historian and Genealogist, I have volunteered to conduct classes for students so that they can learn about and document their family histories. Dr. Felton Williams, chapter Education Chairman, and Mr. Lance Robert, chapter Mentoring Chairman, will monitor the project. We will be working with the students, as well as their families, to get them involved in learning more about their family history and heritage.

Jesse B. Johnson Jr., says, "As we enter the dawn of a new decade, the 100 Black Men of Long Beach, Inc. have positioned themselves, with the help of other community based organizations and public agencies, to contribute to the further development of the Greater Long Beach Community - with emphasis on the African American families, especially the youth."

**Students at the 100 Black Men Gala Event - April 2009
Center – Poster of President, Barack Obama**

Long Beach City College – Sociology 11 – 'Personal Assessment Project'

Janét Hund, Assistant Professor, Sociology

The 2006 & 2007 Sociology Classes at Long Beach City College in Long Beach, California, were assigned a 'Personal Assessment Project'. The teacher was Janét Hund, Assistant Professor, Sociology. As the Genealogist of the project, I presented lectures about 'Family History Preservation' to the classes, (See Appendix III). The students prepared Family History papers, Pedigree Ancestor Charts, and Family Group sheets for the project. Although the students in this project were a little older than those in the NAACP/Jefferson Middle School project, the steps for family history preservation were pretty much the same, **(See Appendix III)**.

***   I told the students to begin by collecting photos, school records, family papers, birth records, obituaries, marriage records, and other documents. I also suggested that they interview their older relatives.**

***   With the assistance of their parents and other relatives, I suggested that they begin researching at libraries, and the counties where their ancestors had lived. They will be able to search through census schedules at the National Archives, the Family History Centers, and other Genealogical Institutions for information about the history of their ancestors. The internet is also available for research on family histories.**

***   To preserve this information for other family members and their descendants, I suggested that they begin putting this data into acid-free sheet protectors and binders.**

***   I showed them how to begin filling out pedigree/ancestor charts, family group sheets, and to start writing about their personal experiences. I also gave them a 15-step plan that they could use through the years.**

TEACHING OUR OLDER GENERATIONS

Preserving the history and heritage of African Americans. In 1997, fifty concerned citizens joined forces to form **The African American Heritage Society of Long Beach (AAHS-LB).** Thirteen of them formed a Board of Directors that would go on to lead the society. With the dedication of these thirteen board members, and the support of the additional charter members, they established the African American Resource Area at Burnett Library to preserve the history of African-Americans throughout the world. They started seven book collections - over 3,000 books.

Annual Programs were presented about African-American History and Heritage, and a Veterans Scroll of Honor was presented as a tribute to 31 Veterans by their families. The group co-sponsored history and cultural programs with other organizations, and genealogy classes and programs were given to help others learn about, and preserve their histories.

In January, 2001, at the October 13th annual meeting, A Grand Celebration of Family, featured history professor Craig Patterson, who talked about his family research, archivist Bill Doty, from the National Archives, who talked about what's available for research, and I lectured about my family history research.

January 5, 2002 was the kickoff for the first AAHS-LB genealogy class with almost 30 participants. As the Chairman of the newly created AAHS-LB Genealogy Council, I designed the beginner's course for the group. Beginning classes were held monthly, at the Burnett Library in Long Beach.

The classes lasted from January to September, and class members were taken to numerous research facilities. With the assistance of Supervisor Don Knabe's office, we received a bus for the day and took a group to the National Archives in Laguna Niguel, California to do research. On other weekends, we took groups to the Los Angeles Central Library, the LDS Family History Center in Los Angeles, the Long Beach Main Library, and the Huntington Beach Library to do research.

As the classes grew, we offered all NAACP-Long Beach Members to join, and participate. We also included a session on Computer/internet and genealogical research that Claudine Burnett, of the Long Beach Public Library started in 2002. She held classes for our group several years, and later, Stephanie Spika and Evelyn Matzat of the Main Library in Long Beach, conducted classes for our society.

In February 2002, the AAHS-LB co-sponsored the first Discover Your Roots: African American Family History Conference with the Church of Jesus Christ of Latter Day Saints at their facility in Los Angeles, along with the California African American Genealogical Society of Los Angeles. We also co-sponsored the first 'West Coast African American Genealogy Summit', with the California African American Genealogical Society, of Los Angeles.

In February 2004, the classes continued, and the AAHS-LB Genealogy Council began participating in various Conferences, so that its members might learn what other groups were doing. The group also began attending the 'Jamboree Genealogy Conference' in Burbank, which is held yearly by the Southern California Genealogical Society.

The society published the best-seller – **'The Heritage of African Americans in Long Beach: Over 100 Years.'** Thanks to this dedicated group of people, the history and heritage of the African American people in Long Beach will never be lost, or forgotten.

In 2007, the AAHS-LB sponsored a 10 - Year Anniversary Program honoring the Society, and six outstanding National Leaders. The group has collected over $35,000.00 to buy books and other materials about African American History and Heritage. This group has made a significant contribution to the preservation and advancement of America's unique African-American History and Heritage.

African American Heritage Society of Long Beach – 13 Original Board Members
Front Row (L-R) Margaret Brown, Phyllis Venable, Maycie Herrington, Doris Topsy-Elvord, Vera J. Mulkey, Evelyn D. Knight, Marie Treadwell, Back Row (L-R) Indira Hale Tucker, Herb Levi, Aaron L. Day, Fannie Holland, Joel Patterson, Marcus O. Tucker, **(All are Charter Members).**

Additional Tips for the Researcher

Writing a family history

After you go through the 15 research steps, you will come to a point where you want to share your information with other family members and friends. This review will explain why family histories need to be preserved, for future generations. The review will also outline some of the necessary steps that need to be taken when writing a family history.

Why do Family Histories need to be preserved?

I suggest that you sit down with your older family members and talk about their history. You may even want to videotape your conversations with them.

A friend of mine invited me to visit, and interview her mother several years ago. We set up a camcorder, and talked to her for over an hour. She told us about her life, family, and friends, and really enjoyed sharing with us.

As it turned out, it was great that we did the interview, because about six months later, she had a stroke. Although, she lived after the stroke, she would not have been able to give us the detailed information that she did at the interview. The following year, she celebrated her 100th birthday, and was surrounded by family and friends.

It is very important that you interview your older relatives and friends as soon as you can, so that you and your families will have the memories.

Why I want to preserve my family history?

History has taken on a new meaning for me. I now realize that my ancestors played a vital part in shaping our nation. To those of us who are researching our family histories, it is very important that we let our younger family members know what we are doing. We must impress upon them that history is also about their families. We must also let them know that someday – they will be a part of this history.

As I research my father's family history, it has been very exciting to locate so many of my Day ancestors. I want to learn as much as I can about each of my ancestors, such as where they lived, how they lived, and what life was like for them. Most importantly, I want to share this information with their descendants. From the notes and records that I have accumulated, I began to detail, and outline the histories and lives of my past ancestors. I decided to document the Day ancestors in book form, and to tell their untold stories. It is my hope, that the Day descendants will have a better understanding of their history and heritage, and will want to preserve both.

Writing your family history

Why do family histories need to be preserved? Writing a Family History is the best legacy that anyone can leave for their descendants. It also helps to preserve your personal and oral history. One great reason to write a family history is that it helps children create a strong bond to their families. It also helps to preserve the history of ancestors. Writing about the present generations, helps to preserve our own lives. Another good reason for preserving a family history is for the future, (our children, and our children's children).

Why do you want to write a family history?

You have to ask yourself a few questions before beginning to write your family history. Are you putting it together just to get the information out to the family so that it won't get lost? Are you trying to put the family history together for the family members who are interested enough to pay for it? Are you hoping to pay for more research? Is your goal to make money?

Steps to be taken when writing a Family History

Read a good book on beginning research and follow the instructions on gathering the material for your book. You will need a lot of Family Group Sheets and Pedigree Charts to make certain that you have all of the information on the family. Make sure you have plenty of storage space to file and maintain your research notes. Having access to a computer will make writing and assembling the book easier. Keep track of expenses as you are writing the book. To begin your research, contact everyone who might be able to help provide information for the book. Organize your filing system so that you can retrieve the information and find the supporting documents. Eventually you will have gathered enough material so that you can begin putting it in the final order that you will be using.

I create a rough draft, or format in pencil or pen. After carefully reviewing this information, my next step is to add it to my computer, and print out a copy. I make corrections from the print-outs, and repeat the process. This works best for me, maybe a little old fashioned. Some people work directly from the computer, and do not write a rough draft. Whatever works best for you, but the main thing is, getting started.

How to present your Family History (Format)

Are you doing just your direct ancestor, or just your descendants? Want to use an ancestral person or couple to use as your starting point? Are you planning to go either further back in time, or to come forward to the present day descendants? Are you going to list all of the information that you have found on one surname? How about mixing families with different surnames together in one volume? You can also present your ancestral information in story-narrative form. Produce a fun book of current family stories. A cookbook family history is another great idea. A newspaper/magazine format is great as a family project, and to include the children. A pictorial history in the form of a scrapbook or heritage album is also a wonderful format.

Sharing your Family History (Publishing)

How do you want the final book to look? Do you want the final book in hardcover or soft cover? Do you want to self-publish?

Review other family histories. Get information from quite a few different publishing companies and compare. When you are comparing costs, list how you want the book to be done, and then compare the costs.

Some helpful sources for writing a family history, and for researching

Writing a Family History by Nancy Ellen Carlberg
Carlberg Press, 1782 Beacon Ave., Anaheim, CA 92804
(714) 772-2849

The Complete Idiot's Guide to – Writing Your Family History
by Lynda Rutledge Stephenson
Macmillan USA, Inc., 201 West 103rd St., Indianapolis, IN 46290

Crafting Your Own Heritage Album
by Bev Kirschner Braun
Betterway Books, 1507 Dana Ave., Cincinnati, OH 45207
(800) 289-0963

Other books and websites of interest

- The Genealogy Sourcebook, by Sharon De Bartolo Carmack
- Climbing the Family Tree, by Nancy Carlberg
- Creative Keepsakes, (888) 247-5282 www.creatingkeepsakes.com
- Creative Continuum, (714) 524-9566 www.creativecontinuum.com
- Heritage Books, Inc. www.HeritageBooks.com
- Everton's Genealogical Helper Magazine, www.everton.com

Book Publishing Information and researching

- Can be found on the Internet by searching under "Book Publishers"
- Genealogy magazines such as; Family Tree, Family Chronicle,
- National Genealogical Society, and Ancestry
- Public Libraries
- LDS Family History Centers
- National Archives

Creating a Timeline - In researching our past family history, we must remember to document and preserve it. We should also document and preserve our present history for future generations.

I recommend creating a timeline or chart for yourself, or whoever you want to write about. Start out by listing several things that happened during each decade of life, starting with the birth up-to the present time. Once you have several things filled in for each decade, begin adding to the list-without going into a lot of detail, and watch the list grow, as you begin remembering memories from the past. You may want to start off with many of the positive things in your life, as I did, and then you can gradually begin filling in some of the challenging experiences that you have faced. The following is an example of the timeline I created for myself, and how I began adding to it. I started with the year that I was born.

Timeline for Aaron L. Day

1939 – Born in Xenia, Ohio.

1940 – Early Life in Ohio – attended Lincoln Grade School.

1952 – Joined East Main Street Christian Church.

1956 – Editor of High School Newspaper, Junior and Senior years.

1957 – Co-editor of Senior Yearbook, -The Eastonian, – (First book).

1957 – Graduation from East High School, Xenia, Ohio.

1960 – Married to Dorothy Lindsay, in the city of Xenia, Ohio.

1961 – Birth of daughter, Carla Day, Xenia, Ohio.

1964 – Graduated from Dayton Technical Institute, Dayton, Ohio (Business Machines).

1964 – Moved to Los Angeles, California (24 years old).

1965 – Employed at Sunset House Mail Order Co. (Computer Oper./Instr.).

1970 – Employed at Security Pacific National Bank – (Bookkeeping).

1975 – Received AA in Accounting from Los Angeles Trade – Technical College, and later received a BS in Business Administration from the University of Phoenix.

1978 – Personal Trust Course, received certificate.

1980 – Relocated to San Diego for 1 year Security Pacific National Bank (became first African American in Administration in the Fiduciary Services Department).

1981 – Submitted recipes to Banks Family Cookbook (published in 1982 honoring great grandparents George and Eliza Banks).

1982 – "Banks" Family Reunion held in Ohio – Celebrated the memory of my Great-Grandparents George and Eliza Banks

1982 – Employed at Alpine Electronics – (Accounting).

1983 – Dealer Award of Excellence from Alpine Electronics, received certificate.

1984 – Academy on Computers, received certificate.

1988 – Employed at Ultramar Refinery Co. – (Accounting).

1989 – The Berlitz Language Centers (Spanish) received certificate.

1990 – Long Beach Project Read Literacy Tutoring, received certificate.

1990 – Volunteer work for Library began (Literacy Tutor).

1990 – "Friends of the Long Beach Public Library" (Board member).

1990 – Friends of the Long Beach Public Library (Book Sales Co-Chair).

1992 – Honored at Literacy Luncheon – Long Beach Public Library (Literacy Tutor).

1992 – Received 2 Letters of Commendation from First Lady 'Barbara Bush' (Literacy Tutor).

1992 – Co-hosted Banks Family Reunion – Los Angeles, California.

1994 – Co-founded "Speakers Bureau" (Project Read Literacy Program).

1995 – Joined Ultramar Community Outreach Committee.

1995 – Received – JCPenney Golden Rule Award (Volunteer Service in Local Community).

1995 – Honored at Project Read 'Tenth Anniversary Celebration' (Literacy Tutor).

1995 – Friends of the Long Beach Public Library formed Community Outreach Committee (Chairperson).

1995 – Began writing column for 'Friends' Newsletter.

1995 – Co-Founded 'Junior Friends of the Long Beach Public Library' With Margaret Durnin.

1995 – Co-founded "South East Neighborhood Association" in North Long Beach – Ninth District (Vice President).

1996 – Co-founded "English/Spanish Connection" Club.

1996 – City of Los Angeles Library Adult Reading Project (Tutor Certificate).

1996 – City of Long Beach-Department of Parks Recreation and Marine (Recognized for North Long Beach Community Parks Program Volunteer Participation).

1996 – Jr. Friends of the Long Beach Public Library – (received City of Long Beach Annual Pride of Partnership Award).

1997 – The 'Breakers' of Long Beach (recognized for Community Volunteer Services).

1998 – "Banks" Family Reunion – African American Civil War Memorial was established in Washington, D. C. (Great-great-grandfather Oscar Daniel Banks served in 1863 – his name is on the memorial).

1998 – Began researching father's family history (Day family).

1999 – Received the Mary Dell Butler 'Volunteer of the Year Award' (City of Long Beach Volunteerism – 1st African American male).

1999 – Jr. Friends of the Long Beach Public Library – (received City of Long Beach Annual Pride of Partnership Award).

2000 – City of Long Beach – Sixth Council District (Community Service-Certificate of Appreciation).

2000 – Began writing column for Questing Heirs Newsletter.

2000 – Began lecturing about Genealogy (National Archives – Laguna Niguel, California).

2000 – Received 2nd place in National Writing Contest, "Tracing My African American Ancestors," (short story).

2000 – City Council approved appointment, by Mayor Beverly O'Neill (Became Recreation Commissioner).

2000 – Recognized as one of fifty Charter Members of the 'African American Heritage Society of Long Beach'.

2001 – Received the 'Older Americans Recognition Award' from the Los Angeles County Board of Supervisors and the Los Angeles County Commission on Aging (Community Volunteerism).

2001 – Founded African American Genealogical Council – (Genealogy classes/workshops for African American Heritage Society members).

2001 – Received 2nd place in National Writing Contest, "The Search For Free African American Ancestors," (essay).

2001 – Questing Heirs Genealogical Society (Vice President – first African American).

2001 – California African American Genealogy Society (became member).

2001 – Published article "The Miniature Tea Set" in Everton's Genealogical Helper magazine.

2002 – Two articles "Dear Aunt," and "History Lessons" published in Gatherings Literary anthology book.

2002 – Received Honorable Mention award in National Writing Contest, "History Lessons," (essay).

2002 – Began writing column in NAACP Long Beach Newsletter

2002 – Received Community Hero Award Long Beach (Volunteerism).

2002 – African American Genealogical Council – (Genealogy classes and workshops offered to NAACP members and also African American Heritage Society members).

2003 – Producer/Director/Host of "Anebel: Still Living the good Life" – (Video).

2003 – Published Genealogy/Reference book -Locating Free African American Ancestors: A Beginner's Guide, (Second Book).

2003 – Vice President – African American Heritage Society of Long Beach 2003 – Published "History Lessons" article in 'Celebrations of Honor' anthology book.

2003 – Published "The Search for Free African American Ancestors" in Heritage Quest magazine.

2003 – Executive Producer "Remember" video honoring Kenneth Slaybaugh).

2004 – Co-Producer/Director of "Explore Your Roots #1 & 2" Genealogy – 2 Videos.

2004 – Host and Co-executive Producer of "Legacy: Sharing our Past" Television Show

2004 – Host, Producer/Director of "Summit On African American Genealogy" (4 videos on African American Workshops-Los Angeles, California).

2004 – Published article "African American Research Resources" in the 'Searcher' Journal of the Southern California Genealogical Society.

2004 – Helped establish "Roots Collection of African American Genealogy" at Main Library Long Beach.

2004 – Received 2nd place in National Writing Contest, "What's Available for African American Research," (advice & how-to).

2005 – Published -History Lessons book, includes poems, essays, and short stories, (Third book).

2005 – Co-Producer/Director of "Explore Your Roots #3" Genealogy, 1 Video.

2005 – Executive Producer and Host of "Marjorie Jonas Schnapp" History Video.

2005 – 1st Annual Congressional "Community Pride Awards," received award from Congresswoman Juanita Millender McDonald.

2005 – "Tracing My African American Ancestors" article appeared in Anthology book called 'Celebrating Family History.

2006 – Published article "Researching Free African American Ancestors – Pre-1865" in the "Searcher' Journal of the Southern California Genealogical Society.

2006 – Published article "DNA to Africa" in Everton's Genealogical Helper magazine.

2006 – Received "Men of Valor" Award from the NAACP Long Beach Branch.

2007 – Co-editor of The Heritage of African Americans in Long Beach: Over 100 Years book published in association with 'The African American Heritage Society of Long Beach,' (Fourth book).

2008 – Became the Historian for the Mombasa/Long Beach Sister Cities Society.

2008 – Published article "Mary Dell Butler" in the 'African American National Biography" book, by Henry Louis Gates Jr., and Evelyn Brooks Higginbotham.

2008 – History Cards – (co-editor Kate Pedigo) set of 15 cards, drawings are by Pedigo, stories by Day, (2008).

2009 – Became the Historian for the '100 Black Men of Long Beach, Inc.'

2009 – Co-founded the "Black Authors Festival Day in Long Beach" with Sunny Nash, and others.

2010 – Published 'DNA To Africa: The Search Continues.

2010 – Wrote the Foreword for "African Americans In Hawaii: A Search for Identity" by Dr. Ayin Adams, PhD.

2010 – Received NAACP Long Beach Branch – President's Award

2011 – Published "Preserving Our History: African Americans in Xenia, Ohio" co-editor Theodore Day.

It is very important that we research, document, and preserve our family histories. I must emphasize, **documenting our own histories for future generations.** My grandchildren are all learning about their family heritage. I have been keeping my daughter, and other family members updated, by periodically sending them reports of my research progress. I don't know if they are bored yet, but I try to keep my stories to them as interesting as possible. I want my grandchildren to know that their family heritage is a part of - the Nation's History.

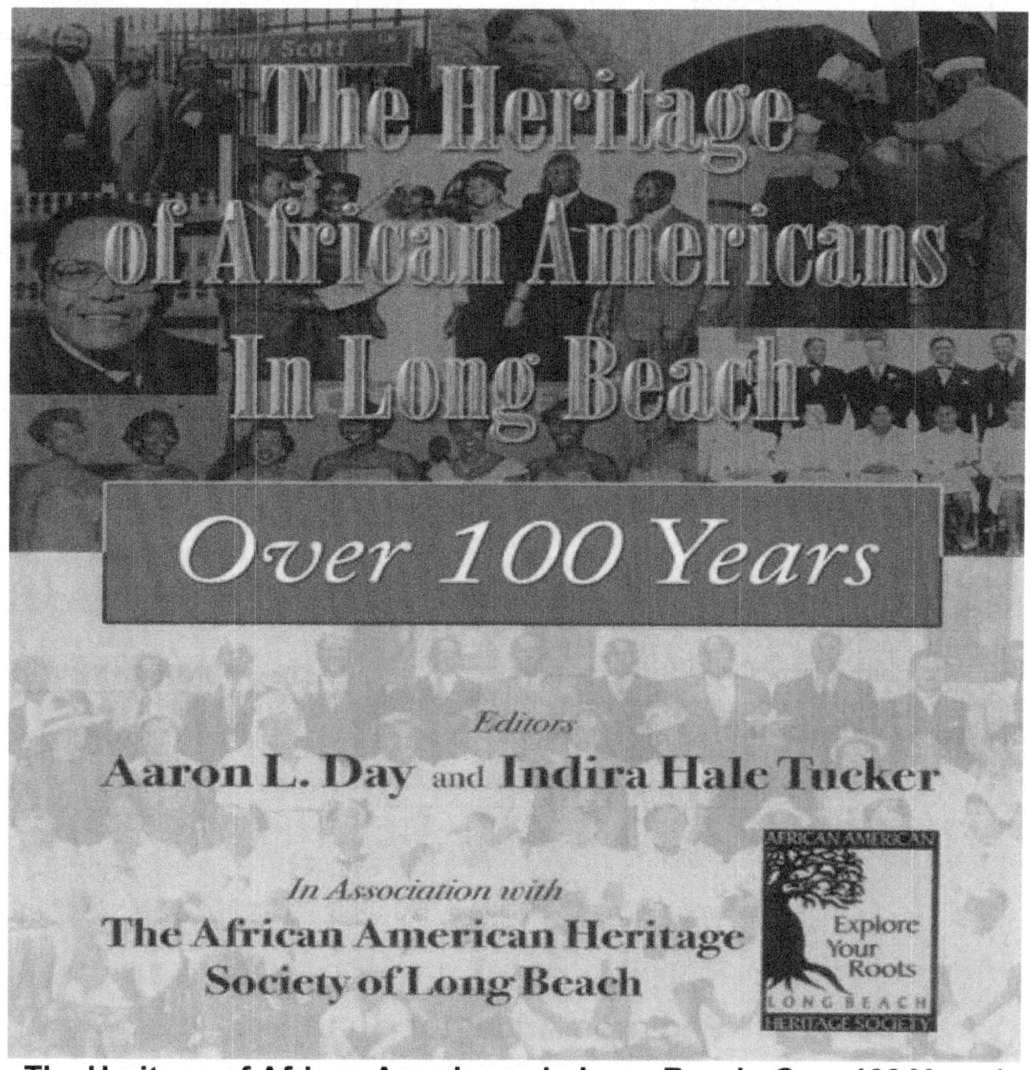

The Heritage of African Americans In Long Beach: Over 100 Years'
InfinityPublishing.com

Aaron L. DAY
Chart 1, generations 1 through 4

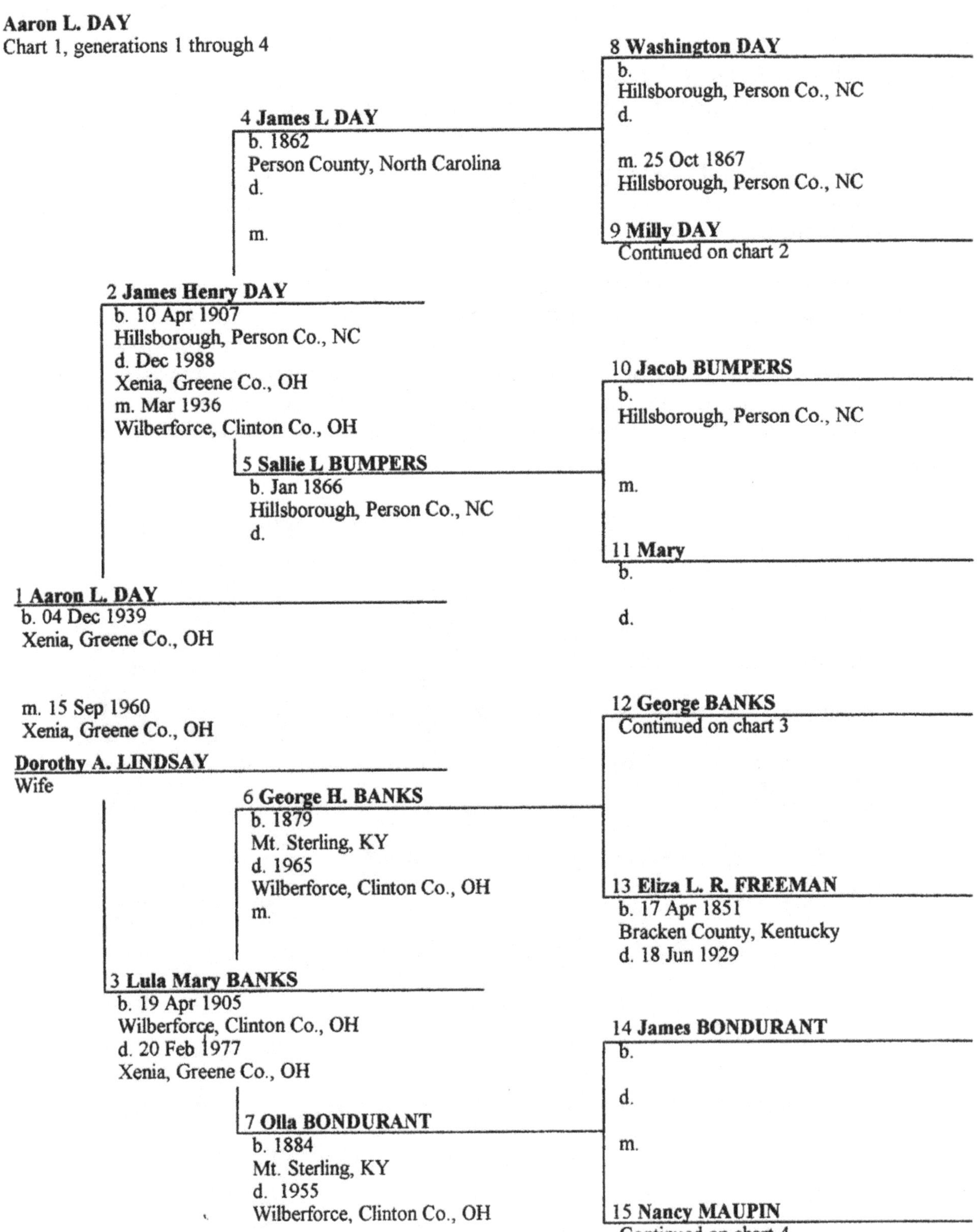

Pedigree Ancestor Chart

CHAPTER 11

HOW TO LEARN MORE

ABOUT YOUR ANCESTORS

Books by Nancy Carlberg

Nancy Carlberg is a former librarian, author of over 40 books, lecturer and genealogy teacher throughout Los Angeles, Orange Counties, and the United States. Nancy was a researcher for Alex Haley, during the **Roots** project. She published Volume one of **Locating Free African American Ancestors: A Beginner's Guide**, and I thank her very much for her contributions to me. Nancy's books have been very helpful to me in learning more about the field of genealogy, especially **Climbing the Family Tree with Nancy.** Nancy's books are listed below.

1. 200 Hints for Genealogist - 2.00

2. Analyzing a Family Group Sheet - 5.00

3. Becoming a Professional Genealogist - 15.00

4. Beginning Census Research - 15.00

5. Beginning Danish Research - 20.00

6. Beginning English Research - 20.00

7. Beginning Irish Research - 20.00

8. Beginning Mexican Research - 20.00

9. Beginning Norwegian Research - 15.00

10. Beginning Scottish Research - 25.00

11. Beginning Swedish Research - 15.00

12. Cemetery Research - 10.00

13. Climbing the Family Tree with Nancy - 15.00

14. Courthouse Research - 3.00

15. Crash Course in French Research - 5.00

16. Dewey Decimal Place List - 10.00

17. Dewey Decimals of the World - 3.00

18. Differences Between Records of the United States, England and Wales, Scotland, and Ireland - 3.00

19. Family Secrets and Scandals - 3.00

20. Genealogist's Guide to Belfast - 10.00

21. Genealogist's Guide to Dublin - 10.00

22. Getting a Quick Start Up Your Family Tree - 30 Minutes At a Time - 10.00

23. How to Survive the Genealogy Bug without Going Broke - 4.00

24. Midwest Research - 5.00

25. Names, Nicknames, and Misspelled Names - 30.00

26. Nancy's Easy Filing System - 5.00

27. Newspaper Research - 3.00

28. No Death Certificate? Now What? - 3.00

29. Overcoming Dead Ends - 20.00

30. Preparing in Advance to Visit a Genealogical Library - 5.00

31. Preserving Your Family Heritage - 5.00

32. Researching in Salt Lake City - 20.00

33. Pet Rock Pedigree Chart - 2.00

34. Teaching Genealogy -25.00

35. Toy Making on a Budget - 3.00

36. Tracing Your Colonial Ancestors - 25.00

37. Travel Hints for the British Isles - 3.00

38. Traveling to England to Find Your Roots (Cheap) - 20.00

39. Traveling to Ireland to Find Your Roots - 20.00

40. Using the 1992 IGI on Microfiche - 3.00

41. Writing a Family History - 10.00

42. Writing a Non-Fiction Book the Easy Way - 10.00

Books by other authors:

1. Locating Free African American Ancestors: A Beginner's Guide - 15.00 & 20.00

Books by other companies available through Carlberg Press:

1. Beginning Welsh Research - 20.00

2. Bibliography of Welsh Sources - 8.00

3. Planning a Genealogical Research Trip to Wales - 15.00

4. Techniques to Finding the Place of Origin of Your Welsh Ancestor - 3.00

5. Welsh Places and Farm names - 12.00

6. Welsh Surnames and Given Names - 15.00

Children Buried in Iowa Cemeteries:

7. Buchanan Co. - 5.00

8. Appanoose Co. - 4.00

9. Adair Co. - 3.00

10. Johnson Co. (excluding Iowa City) - 5.00

11. Lucas Co. - 5.00

12. Cass Co. - 5.00

Military Buried in Iowa Cemeteries:

13. Buchanan Co. - 5.00

14. Appanoose Co. - 3.00

15. Adair Co. - 3.00

16. Johnson Co. (excluding Iowa City) - 5.00

17. Lucas Co. - 3.00

18. Cass Co. - 3.00

Carlberg Press, 1782 Beacon Avenue, Anaheim, CA 92804-4515

Nancy Carlberg, also lectures at the Annual

'Discover Your Roots' Conferences

African American Genealogy

Thanks to Alex Haley, creator of the *Roots* saga, people from all walks of life have developed an interest in family history research. Discovering information about ones heritage can be very rewarding. For many African Americans, however, the search can become quite challenging because of the institution of slavery.

Fortunately for us, there are historians who have followed in Haley's footsteps and have documented their research methods and procedures. Over the past 30 years since *Roots* first appeared countless articles and books have been published to help researchers who are looking for information about their ancestors. For those who want to learn more about genealogy, there are numerous categories of books and records to choose from. Some of the many categories of books on genealogy and family history preservation are listed below.

CATEGORIES:

1. Beginning guides for research

2. Intermediate/advanced research

3. How-to books on genealogy

4. Books on slave research

5. Books on researching free ancestors before 1865

6. Internet resources

7. Public records, including

- **Census records**
- **Military research**
- **Birth**
- **Death**
- **Marriages**
- **Deeds/land records**
- **Court records**

8. Additional research records including
- **Church records**
- **Cemetery research**
- **Family books**
- **Scrapbooks/family cookbooks**
- **Family reunion records**

As you can see from these various categories, genealogy/family history research covers a very wide range of subjects.

Like those in other professions, many genealogists or family historians have begun to specialize in certain areas. What's available for African American research? A growing number of genealogy books are geared specifically for African American research. The historians publishing these books have put considerable time and effort into sharing and passing on their discoveries, to others who are researching ancestors.

A recent review by the author for the Southern California Genealogical Society on what is available for African American research, revealed the following:

Family Pride: The Complete Guide to Tracing African-American Genealogy, **by Donna Beasley; Publisher: Hungry Minds Inc (February, 1997) ISBN: 0028608429**

In *Family Pride*, Ms. Beasley gives step-by-step instructions on how to research African American family history and genealogy. Her chapters include information on how to begin, oral history, slavery, the African connection, writing and publishing your findings, and working with technology. In her bibliography, she lists books and periodicals that best support genealogy research.

Black Roots: A Beginner's Guide to Tracing the African American Family Tree, **by Tony Burroughs; Publisher: Fireside (February 6, 2001) ISBN: 0684847043**

In *Black Roots*, Mr. Burroughs delivers a step-by-step "how to" book to help those starting to research African American family history. He uses many of the methods used in an introductory genealogy class that he teaches at Chicago State University. His chapters include information on preparing to research, which covers the fundamentals of getting started. It also features guidelines on what types of records are available for research and where to find them, samples of worksheets and forms needed for research, using computers and the Internet for research, and writing a family history.

A Genealogist's Guide to Discovering your African-American Ancestors, **by Franklin Carter Smith and Emily Annie Croom; Publisher: Betterway Books (2003) ISBN: 1-55870-605-4**

In *A Genealogist's Guide*, Mr. Carter and Ms. Croom provide step-by-step techniques for research that may be useful to those who want to learn more about their ancestors. Their chapters include information on basic principles of genealogy, census records, federal resources, state-county-local sources, a study of names (given names and surnames), slaveholder documents, the issue of mixed race, and special situations, such as free Negroes before the Civil War.

Finding A Place Called Home: A Guide to African-American Genealogy and Historical Identity, **by Dee Parmer Woodtor; Publisher: Random House Reference; Rev & Expand edition (November 16, 1999)**
ISBN: 037570843X

In *Finding A Place Called Home*, Ms. Woodtor gives us methods for searching and interpreting records. She outlines interviewing techniques for family members and friends, and also shares information about using the Internet for genealogical purposes. Her chapters include information on beginning your genealogical research, techniques and tools used, searching for ancestors during the Civil War era and during Reconstruction, slaves and slave owners. Additional information includes African American institutional records, and what to do with your research: writing family memoirs or the family story.

Tracing African-American Roots,
by Dee Clem; Publisher: Gator Pub; 1st   edition (February, 2000)
ISBN: 0967584612

In *Tracing African-American Roots*, Ms. Clem presents a great how-to book that provides general guidelines on how to use oral history, personal family data, and public records to find your ancestors. Her chapters include information on starting your search for African American ancestors, organizing your records, the importance of libraries and archives, and census schedules. The following records are also reviewed; vital, cemetery, mortuary, church, court, land and tax, military, slavery, and miscellaneous records. Resources on the Internet are also covered.

How To Trace Your African-American Roots,
by Barbara Thompson Howell; Publisher: Citadel Press (January, 1999)
ISBN: 0806520558

In *How to Trace Your African-American Roots*, Ms. Howell produces a practical guide that shows you how to use the basic resources of every genealogist to trace your ancestors and more. Her chapters include information on where to find and what to look for in birth records, marriage certificates, deeds, death records, wills, and census records. For beginning genealogists/family historians Howell outlines how to start, how to keep track of everything, how to locate family documents, and how to investigate church, cemetery, census, and town and county records.

Finding Your African American Ancestors: A Beginner's Guide,
by David T. Thackery; Publisher: Ancestry Publishing (October, 1998)
ISBN: 0916489906

Finding Your African American Ancestors is a compilation of works by the late David T. Thackery. Mr. Thackery was a curator of local and family history at the Newberry Library in Chicago for fifteen years. His chapter in "The Source: A Guidebook of American

Genealogy" as well as his guide to African American research at the Newberry Library is included in this book.

Other chapters in this book include information on free blacks, the Underground Railroad, the transition from slavery to freedom, and military records. Also included are slave narratives from various states, and additional African American sources.

***Black Genesis*, by James Rose and Alice Eichholz; Publisher: Genealogical Publishing Company; 3rd edition (April, 2003) ISBN: 0806317353**

In *Black Genesis*, Mr. Rose and Ms. Eichholz outline suggested steps for the beginning genealogist/family historian. They encourage the development of black genealogical research and also offer ways genealogical materials can be used to re-examine history.

Their chapters include information for the novice – an introduction to genealogy, genealogical guidebooks, and oral history. National Archives and federal records are reviewed, as well as war records – Revolution (1776), the War of 1812, the Civil War, and other military records. There is also a complete chapter on slave research, including such topics as bills of slave sales, slave advertisements, court records, plantation records and diaries, and other records that may be useful.

***Locating Free African American Ancestors: A Beginner's Guide*, by Aaron L. Day; Publisher: Carlberg Press, Anaheim, California (2003) www.banksday.com and www.day-banks.com ISBN 0-944878-52-0**

In *Locating Free African American Ancestors*, the author presents a reference manual that is a great aid for genealogy researchers. It gives tips on problem solving techniques, research methods, and resources for locating ancestors. Included, is information on beginning your search, tracing your ancestors through the census schedules, locating information about free ancestors from documents, and following the paper trail to your ancestors.

The chapters also include information on searching for free ancestors before 1865, genealogical forms, steps for tracing your ancestors, genealogy resources on the Web and your public library, resources available for researching your ancestors, and genealogical books and magazines. There is also a listing of surnames of Negroes who were free in the United States in 1830.

***A Student's Guide To African American Genealogy*, by Anne E. Johnson, Adam Merton Cooper, and Roger Rosen; Publisher: New York: Oryx Press (1996) ISBN: 0-89774-972-3**

In *A Student's Guide to African American Genealogy*, the authors illustrate the unique and important contributions of African Americans to American culture. The book is tailored not only for African Americans who are trying to trace their roots back to Africa, but also for those who may be interested in African American family history research.

The first chapter of this book lists other books for those interested in further study. The chapter 'Starting Your Exploration' details fourteen books on African American History, and 'African American Language and Culture' reviews 45 books on African American History. There are also eleven books of interest on African language and culture. Other chapters include information about slave life, free blacks and freedmen, and reconstruction and modern history. For the beginning researcher, there is also information on getting started, where to research, family history on the Internet, and preserving your family history, which includes the topics family tree, oral history, and writing your family history.

Black Family Research: Records of Post-Civil War Federal Agencies at the National Archives, **by Reginald Washington; Publisher: National Archives and Records Administration – Reference Information Paper 108;**
http://www.archives.gov/publications/finding-aids/guides.html#blackhist.

In *Black Family Research*, Mr. Washington reviews some of the most important records available for the study of black family life and genealogy. These are Reconstruction era federal records that document the black family's struggle for freedom and equality. These documents are available for research at the National Archives and Records Administration.

This reference booklet describes three post-Civil War federal agencies' records. (1). The Bureau of Refugees, Freedmen and Abandoned Lands (Record Group 105), also known as the Freedmen's Bureau, which was established in the War Department by an act of Congress on March 3, 1865. (2). Records of the Freedman's Savings and Trust Company, which was established as a banking institution primarily for the benefit of former slaves. (3). Records of the Commissioners of Claims, which was established by an act of Congress on March 3, 1871, to review and make recommendations regarding the claims of Southern Loyalists who had "furnished stores and supplies for the use of the Army" during the Civil War.

Free Negro Heads of Families in the United States in 1830, **by Carter G. Woodson; Published by the Association For The Study of Life and History, Inc. (1925), and Publisher: University Microfilms International (1979) ASIN: B000878WBY**

Black and Free: The Free Negro in America, 1830, edited by, Alan Abrams, with a commentary on Carter Woodson's "Free Negro Heads of Families in the United States in 1830" Publisher: Doubting Thomas Publishing (2001) ISBN 0-9714303-0-6

In *Free Negro Heads of Families*, Mr. Woodson documents the study of the Free Negro in the United States 1830 and beyond. It was his aim to promote the further study of a part of our history that had been neglected. With a grant from the Laura Spelman Rockefeller Memorial in 1921, Mr. Woodson and a staff of researchers documented the Negroes in this country who were free prior to the Emancipation in 1865. Written in 1925, the information in this book was extracted from the manuscript schedules of the 1830 United States Census.

In the introduction, Mr. Woodson covers the origin of the free Negro, the attempt to prevent the increase of the free Negroes, economic achievement, the free Negro before the law, and social distinctions. He also references other books of interest that have been written on the subject. There were 319,498 free Negroes in the United States in 1830, and they were living in 28 states. The surnames of those families are listed in this book, as well as an index to the names, and the states where they were living.

Searching for free African American Ancestors pre - 1865

The powerful saga *Roots* by Alex Haley had a tremendous influence on people throughout the nation. As a result of the book, and the movie, people wanted to learn more about their ancestors. African Americans in particular were deeply affected and inspired by Haley's ability to trace his family back to Africa.

Since Roots was released over thirty years ago, historians who have done family history research have been documenting and publishing their findings. Much of this information deals with the free black population living in the United States before the end of the Civil War in 1865. There were over 300,000 free blacks living in the United States in 1830, and they were living in twenty-eight states. That number had increased to over 500,000 by 1865 when the Civil War ended. You can imagine how many descendants they have today.

After discovering that some of my father's ancestors had been free before 1865, this author began learning more about that segment of the population. Historians such as Carter G. Woodson, Leon F. Litwack, Ira Berlin, Leonard P. Curry, and John Hope Franklin, to name a few, have published exceptional books about free African Americans who lived before 1865.

There have been a great number of books written about free African Americans who lived in various states, cities, and counties. Some of these books offer information about social conditions, free registers, manumissions, special laws, educational opportunities, and more. A few of these books are listed below.

- *The Free Negro in Virginia 1619-1865.*
- *The Free Negro in Maryland 1634-1860.*
- *The Free Negro Heads of Families in The United States in 1830.*
- *North Of Slavery: The Negro in The Free States 1790-1860.*
- *Northumberland County Virginia Registers of Free Blacks.*
- *The Free Black in Urban America, 1800-1850: The Shadow Of The Dream.*

- *The Free Negro in North Carolina 1790-1860.*
- *District Of Columbia Free Negro Registers 1821-1861.*
- *Free African Americans Of North Carolina & Virginia/Maryland & Delaware.*
- *Slaves Without Masters: The Free Negro in the Antebellum South.*
- *Free Blacks and Mulattos in South Carolina 1830 Census.*
- *Locating Free African American Ancestors: A Beginner's Guide.*

Many of these books actually list the names of free blacks that were living during that time period. These books were compiled from information taken from a variety of documents such as census schedules, free register records, manumission papers, court records, land entry records, and other sources.

The historians who have written and published these books are to be commended. They have spent considerable time and effort to pass this information on to other researchers who are interested in learning more about the status and the history of the free black population before 1865. A review of many of these books by the author revealed the following.

The Free Negro In Virginia 1619-1865,
by John H. Russell; Publisher: Johns Hopkins Press; (1913)
ASIN: B00085W98S

In *The Free Negro*, Mr. Russell explores the status of the free Negro in the slave state of Virginia. His chapters include information about the number and distribution of the free Negroes, manumission, and the origin of the free Negro class. Russell reviews some of the popular misconceptions about the beginnings of the free Negro population in Virginia. Russell also analyzes the five classes of colored persons that increased the free population, which were; children born of free colored parents, mulatto children born of free colored mothers, mulatto children born of white servants or free women, manumitted slaves, and children of free Negro and Indian mixed parentage.

The last two chapters of the book discuss the legal status of the free Negro, and the social status of the free Negro. The primary sources for this book are listed in the bibliography. These sources include manuscripts, laws and court decisions, public documents, newspapers, magazines and periodicals, published parish records and histories, and contemporary works and pamphlets. Russell published his book in 1913, and was a pioneer in advocating the study of the history of free Negroes.

The Free Negro In Maryland 1634-1860,
by James M. Wright; Publisher: Columbia University; (1921)
ASIN: B00085X3ZQ

In *The Free Negro In Maryland*, Mr. Wright talks about the development of this class of people during the early growth of Maryland. He also gives the reader information about several earlier works on free Negroes. Two of these works are *White Servitude In Maryland* by Mac Cormac, and *The Negro In Maryland* by Jeffrey R. Brackett. The chapters include information about the colonial period, the legal status of the free Negroes, and the apprenticeship of Negro children. In addition, Wright talks about social conditions, educational opportunities, churches, and the occupations of the free Negroes in Maryland.

In the chapter "Growth of the Free Negro Population," Wright presents a table that outlines the free colored population of Maryland, 1790-1860. They were living in 21 counties. The total grew from 8,043 in 1790 to 83,942 in 1860. There are also tables showing wages paid in certain counties, and a large number of property holders who were living in various counties. Wright includes an extensive bibliography that will surely be a great aid to researchers who are interested in the Maryland area.

***Free Negro Heads of Families in the United States in 1830*,
by Carter G. Woodson; Published by the Association For The Study of Life and History, Inc. (1925), and Publisher: University Microfilms International (1979) ASIN: B000878WBY**

***Black and Free: The Free Negro in America, 1830, edited by, Alan Abrams, with a commentary on Carter Woodson's "Free Negro Heads of Families in the United States in 1830" Publisher: Doubting Thomas Publishing (2001)
ISBN 0-9714303-0-6***

In *The Free Negro Heads of Families*, Mr. Woodson documents the study of the free Negro in the United States in 1830 and beyond. It was his aim to promote the further study of a part of our history that had been neglected. With a grant from the Laura Spelman Rockefeller Memorial in 1921, Mr. Woodson and a staff of researchers documented the Negroes in this country that were free prior to the Emancipation in 1865.

Extracted from the manuscript schedules of the 1830 United States Census, this book also covers the origin of the free Negro, the attempt to prevent the increase of the free Negroes, economic achievement, the free Negro before the law, and social distinctions. Woodson also references other books of interest that have been written on the subject. There were 319,498 free Negroes in the United States in 1830, and they were living in 28 states. The surnames of those families are listed in this book, as well as an index to the names, and the states where they were living.

***North Of Slavery: The Negro In The Free States 1790-1860*, by Leon F. Litwack; Publisher: The University of Chicago Press; (1969)
ASIN: B000B9MBEI**

In *North Of Slavery*, Mr. Litwack compares the status of Negroes who lived in the North to those who were living in the South during the antebellum years. He points out that the Northern Negroes did have one great advantage over the Southern Negroes, and that was their ability to organize, petition, and push for improvement of their condition. In the chapter "From Slavery To Freedom," Litwack reviews the gradual abolition of slavery in the Northern States, and also the colonization movement. Chapters on the Federal

Government and the free Negro, and the politics of repression focus on the federal rights and privileges of the free Northern Negro, as well as some of the repressive laws that were put into effect to deny certain rights.

Litwack also compares the educational system, and the churches in the North, explaining that separate facilities were maintained during that period. However, some Northern schools did admit Negroes, and several religious denominations included churches for whites as well as Negroes. Abolition and the crisis of the 1850's are both covered in this book. For those interested in the study of Black History, Litwack's remarkable bibliographical essay is especially valuable because of the many resources that are named.

Northumberland County, Virginia Registers Of Free Blacks, by Karen Sutton; Publisher: Heritage Books; (1999) ISBN: 0788411322

In *Registers of Free Blacks*, Ms. Sutton gives a brief history of the free Negro, and then lists the Free Negro Registers of Northumberland County, Virginia. In reviewing the history of the free Negro in chapter 1, tables 1 & 2 give a race and status breakdown of the 1790 census for certain Virginia counties. In chapter 2, the Registers of Free Negroes from 1803 to 1858 are given for the county of Northumberland.

It was in 1793, that Virginia's General Assembly passed a law requiring all free blacks and mulattos to go to their local courthouse and register. They in turn were given a registration number and a certificate that was to be carried at all times. The registers listed a person's name, colour, age, height, marks or scars, and date of register. There is also a column that tells if the person was born free, or emancipated. This author was fortunate enough to locate over fifteen of his ancestors in these registers.

The Free Black In Urban America, 1800-1850: The Shadow Of The Dream, by Leonard P. Curry; Publisher: University of Chicago Press; (1981) ISBN: 0226131246

In *The Free Black in Urban America*, Curry explores the free black experience in fifteen cities, for a period of fifty years. In order to do a more diverse study of this group; he chose small cities as well as large cities, and not the fifteen with the largest free black population in 1850. The fifteen cities are Albany, Baltimore, Boston, Brooklyn, Buffalo, Charleston, Cincinnati, Louisville, New Orleans, New York, Philadelphia, Pittsburgh, Providence, St. Louis, and Washington.

Curry talks about the growth and nature of the urban black population, and he notes that unlike the slaves, free persons of color were found in all fifteen cities throughout the first half of the nineteenth century. Curry reveals that the major sources of the continuing growth of the free black population were emancipation, manumission, and natural causes.

Other issues examined are occupations, property ownership, housing, and the residential pattern. Curry also covers urban black education, the Negro church in the city, and community activities. In "Note on Sources," Curry gives an extensive listing of works that other researchers might find useful. In the "List of Initial Citations," Curry thoroughly documents this well written book on the free black population in Urban America.

The Free Negro In North Carolina 1790-1860, by John Hope Franklin; Publisher: University of North Carolina Press; (December, 1995) ISBN: 0807845469

In *The Free Negro*, Mr. Franklin talks about the quality of life, the harsh treatment, and the study of the free black in the South, particularly, North Carolina. The research and writings of Mr. Franklin about free blacks in the early 1940's stimulated an interest in the subject, and others soon followed. In the introduction, Franklin gives credit and praise to the historians who preceded him, and notes their names as well as books and manuscripts written by them. In exploring the social life of the free Negro, Franklin talks about education, religion, and the social relationship of the people.

His chapters include growth of the free Negro population, the legal status of the free Negro, and the free Negro in the economic life of North Carolina. In the chapter on "An Unwanted People," Franklin reviews the growing hostility to free Negroes and the colonization movement. In his excellent bibliography, Franklin lists the numerous resources he used to compile this book.

Primary sources used are; manuscripts, books and pamphlets, newspapers and periodicals, and public documents. Secondary sources used are; books and monograms, and articles. The bibliographic afterword should prove very helpful to anyone studying free blacks that lived before the end of the Civil War. It contains a listing of 50 works on free blacks.

District Of Columbia Free Negro Registers 1821-1861, by Dorothy S. Provine; Publisher: Heritage Books Inc; (July, 1996) ISBN: 0788405063

In *District Of Columbia*, Ms. Provine has compiled the abstracts of entries in the registers for free blacks for the District of Columbia. There were originally 5 volumes of free register records for the period between 1821 to 1861, however, volume number 4 covering the period 1846-1855 is now missing. There are approximately 3,600 persons included in the four remaining volumes. The registers provide brief physical descriptions of the person, name, and sometimes information regarding a relative.

In the introduction, the author gives several different reasons why a person may have been listed as free. For further study, Provine includes a listing of books with information about free African Americans in the antebellum period. In compiling and publishing the abstracts, Provine has provided an index with names and the registration numbers so that persons can be easily located in the text.

Free African Americans of North Carolina, Virginia, and South Carolina from the Colonial Period to About 1820 by Paul Heinegg; Publisher: Clearfield (January, 2001) **ISBN: 080635111X**

Free African Americans of Maryland and Delaware from the Colonial Period to 1810 **by Paul Heinegg; Publisher: Clearfield Co (2000) ISBN: 0806350423**

In *Free African Americans*, which was published in 1997, Mr. Heinegg gives the genealogies of 281 free African American families who were living in Virginia and North Carolina during colonial times. Heinegg concludes that "most of the free African Americans of Virginia and North Carolina originated in Virginia where they became free in the seventeenth and eighteenth century before chattel slavery and racism fully developed in the United States."

In the introduction, Heinegg tells of some of the first African slaves who became free, African Americans who were taxable in the 1670's, where the free populations lived, and the white, black, and mulatto connection. Also examined are the discriminatory tax laws, children of the free who were bound as apprentices, the stealing of free people, the "Free Negro Code" laws, and the Native American Influence.

This author located his 3 great-grandfather, and his 4 great-grandmother and many more ancestors in this book by Heinegg. In the year 2000, Heinegg published "Free African Americans of Maryland and Delaware," from the colonial period to 1810.

Slaves Without Masters: The Free Negro In The Antebellum South, **by Ira Berlin; Publisher: Random House Inc (T); 1st edition; (December, 1974) ISBN: 039449041X**

In *Slaves Without Masters*, Mr. Berlin relates how free Negroes lived, and what type of work was available for them during the early years of the country. He also describes how they were able to educate, entertain, and protect themselves.

His book is divided into three parts. Part one deals with the emergence of the free Negro caste, 1775-1812. The chapters cover the origins of the free Negro caste, from slavery to freedom, the failure of freedom, and the free people of color of Louisiana and the gulf ports. Part two talks about the pattern of free Negro life.

The chapters include information about patterns of growth, white racial attitudes and policies and the free Negro community. In part three, Berlin reviews the crisis of the 1850's. This part talks about the best of times, and the worst of times, with an epilogue discussing freemen and freedmen. There are several charts in Appendix 1 that detail the slave and white populations in sixteen states. In Appendix 2, Berlin lists the extensive manuscript sources consulted for the preparation of this excellent book that was published in 1974.

***Free Blacks and Mulattos In South Carolina 1850 Census,* by Margaret Peckham
Motes; Publisher: Clearfield Co; (October, 2000)
ISBN: 0806350261**

In *Free Blacks and Mulattos In South Carolina*, Ms. Motes abstracts the free blacks and
mulattos from the 1850 South Carolina Federal Census. There are over 8,000 free black
and mulatto names listed in this book. They are between the ages of 1 month to 112 years.
There were 29 counties in the 1850 South Carolina Federal County Census that were
abstracted for this review. Motes states that "during this study, thirteen reels of microcopy
were reviewed." Persons are listed by first and last names, ages, sex, and occupations
when known. A person's birthplace, as well as color – black, mulatto or white is also listed.
Also included, are a name index, an occupation index, and a place of birth index that helps
researchers locate their ancestors easier.

***Locating Free African American Ancestors: A Beginner's Guide*,
by Aaron L. Day; Publisher: Carlberg Press, Anaheim, California (2003)
www.banksday.com and www.day-banks.com ISBN 0-944878-52-0**

In *Locating Free African American Ancestors*, the author presents a reference manual that
is a great aid for genealogy researchers. It gives tips on problem solving techniques,
research methods, and resources for locating ancestors. The author includes information
on beginning your search, tracing your ancestors through the census schedules, and
locating information about free ancestors. Chapters include information on searching for
free ancestors before 1865, genealogical forms, steps for tracing your ancestors,
genealogy resources on the Web and your Public Library, where to look for information,
and genealogical books and magazines.

Marleta Childs of the Amarillo Globe-News writes in her, Kin-searching column "perhaps
the most important thing about 'Locating Free African American Ancestors' is that the book
makes more people in general, and more family researchers in particular, aware of the
wide variety of resources available on the subject." There is also a listing of surnames of
Negroes who were free in the United States in 1830.

Good luck in finding more information about your ancestors. In searching for your
ancestors, you will also be sure to learn more about the history and development of our
country. As you continue your research, be sure to document and preserve your findings
for future generations. Remember, you are the link to the past and to the future.

EPILOGUE

In a few weeks, when I receive the results for my paternal grandmother, I will possibly have the ancestry of all four of my grandparents. The link between DNA and genealogical research has progressed tremendously over the past ten years. I am very happy that I heard about the Molecular Genealogy Research Project and that I have been able to make the connection to my African Roots. Many well known people have gone through the DNA testing procedure and have traced their roots back to Africa. I have talked to friends who have discovered their African ancestry through DNA testing. Some of them have already made the trip to Africa, and others are still in the planning stages. They all feel that the discovery of their ancestral homelands is something wonderful to share with family members.

Now I am spending more time at various libraries, and on the Internet trying to learn more about my ancestors' homeland. Members of our family are already making plans for a trip to Africa next year. On the maternal side of our family, (Banks) we usually have a family reunion every year. My cousin Cordelia Thompson Hill, who is normally involved in the planning for these reunions, is the person who is spearheading the effort for the trip to Africa. She is the cousin whose son gave her the DNA Test Kit as a Christmas present.

The search for my free ancestors has been a wonderful experience for me, although painful at times when I learned of the harsh laws that were put into place. The discovery that the Day families were free in the deep-South was quite a surprise to me and my family members. Locating free African American ancestors has also taken me on a long and unexpected journey through the past history of our country. I have learned so much, and I am still continuing to learn.

Daniel Webb & Isabel, Mary Day & Daniel Webb, and Washington Day are people that I am still focusing on. There is still much to learn about the lives of their ancestors - in America, Europe, and Africa. Who knows, maybe another book?

The August 1998 Banks family reunion honoring my great-great-grandfather, Oscar Daniel Banks changed my life tremendously. I now realize the importance of preserving family histories for future generations. With my family members, (Banks and Days) I am now doing my part to learn about our ancestors. This information will be passed on to our descendants so that they will know their heritage. Hopefully, you have been inspired, and will want to learn more about your family heritage, so that you may pass valuable information on to your descendants.

Doing the research for my ancestors has been very exciting for me. I never dreamed that I would be able to find so much information about my family. I have also learned a lot about my personal feelings regarding history. In "I Know Who I am" I talk about different stages of my life – from early childhood to the present. **"I Know Who I am",** is dedicated to the memory of Dr. Martin Luther King, Jr., who continues to be a source of inspiration for many.

"I Know Who I am" is presented here, as a source of inspiration to others.

I Know Who I am

As a child, I grew up,
thinking that I was Colored.
We were called Colored people then,
as that was the popular phrase.

As a teenager, I learned of Dr. King,
and found out that I was Negro.
We were called Negro people then,
as that was the popular phrase.

As I grew into adulthood,
I learned that I was Afro-American.
We were called Afro-American people then,
as that was the popular phrase.

As a young man,
I discovered that I was Black, and proud.
We were called Black people then,
as that was the popular phrase.

As an older man,
I am adjusting to being an African American.
We are called African American people now,
as that is the popular phrase.

As I grow older, and still inspired by Dr. King,
there is one thing that I have learned.
No matter what the popular phrase is,
I know who I am.

By Aaron L. Day

APPENDIX I

Making the connections, and putting it all together

The following are the types of documents that were used to connect the Day ancestors: additional land entries, court records, deed books, etc. Many of the documents are difficult to read, because of the abbreviations, sentence structures, language uses, spellings, punctuation's, and other things.

At times, it was like reading a foreign language. But, like learning a foreign language, I eventually began to decipher, and understand the many documents that are available for family researchers.

FROM Caswell County, North Carolina Land Entries, 1778-1795, on page 103, #598, Aug. 25, 1778. William Day enters 400 AC on both sides of Tarr R; border: William Tapp and William Yarborough; includes his own and Widow Day's improvement {sic.}

FROM the same book, page #828 – Dec. 14, 1778. John Day enters 200 AC on the "Mane" branch of Tarr R; border: William Tapp; includes his improvement.

I interpreted the information from the above entries to discover that, there were three Days living in Caswell County near a William Tapp in 1778. It is possible that the widow Day is the mother of William and John. (A or AC = acres), (Tarr R = river), (Mane = main), (Cr = creek), (adj. = adjoining), (Wit. = witness).

FROM an entry in Deed Book 'A' – 3 other neighbors of William Day and William Tapp. Thomas Person, William Yarbrough, and Burgoon Bird are living near the Tarr R area. Twelve years later, in 1791, this area was renamed Person County, North Carolina for a Thomas Person who fought in the Revolutionary War.

FROM the same Deed Book 'A' page 613 – William Day of Surry Co., NC, to Robert Davis of CC, for 50 lbs, 150 'A; on Tar R adj. William Tapp, Thomas Person, 7 Oct 1784. Wit; James Wilson, Henry Day. Page 115 – Obediah Collins of Surry Co. NC, to Abram Self of CC, for 100 lbs, 300 'A' head of Buck Mountain Cr adj. Thos Person. 6 Jan 1785. Wit; Henry Day, Manley Winstead.

The Day families were found living near Thomas Person, before and after he went to the War. They worked in clearing the brush, paving the roads, and other types of work. These entries both show that there was also another Day in the area (a Henry Day). There was also another Winstead listed (a Manley Winstead). As you can see from the above entries, the grammar and sentence structure was not as well developed as it is today. Different people spelled the same word the way it sounded to them (Tarr and Tar), (Thos = Thomas).

FROM Deed Book 'D' - 5 entries were located containing information about the Days. Page 35, No. 664 – To William Waite 512 'A' on Tar R and Deep Cr adj. John Day, William Day, Thomas Person, Thomas Day, Vincent Harrison, John Clayton. 10 Nov 1784. Page 55, No. 774 – To Tabitha Pryor 640 'A' on Deep Cr and Tar R adj. John Day, Wm Tapp, Robert Dickens, John Thomas, John Bumpass, Hardy Crews. 10 Nov 1784. Page 120, No. 845 – To John Day 200 'A' on Tar R adj. William Tap, William Day, 10 Nov 1784. Page 385-6, No. 865 (Grants from State of North Carolina) – To Thomas Person 500 'A' Middle Prong Tar adj. Campbell, Parker, Day, Davis, Wilson, land Person purchased of Harrison. 5 Nov 1787. Page 387-8, No. 1170 (Grants from State of North Carolina) – To Richard, Stephen, Powell, David, Frances, & Person, Campbell, Tapp, Jones, Wm. Day, Parker old line. 1 Nov 1787. These entries list the names of more neighbors who were in the area during that period. The following three entries from Deed Book 'E' continue to show who is buying and selling land in the area (Deep Cr = Deep Creek), (Wm = William).

FROM Deed Book 'E' page 188 – Samuel Winstead of CC to Cotance Winstead of same, for 100 lbs, 250 'A' on Mayo Cr adj. Gabriel Davey, Drury Allen, land Winstead purchased of State. 10 May 1787. Wit; Gabriel Davey, James Davey. Page 195 – Thomas Person of

Granville Co., NC to Robert Paine of CC, for 150 lbs, 150 'A' being land purchased of William Jay Junr. 4 Aug 1774 and part of his father's original tract on Flat R on Hillsborough Rd. 28 Mar 1787. Wit; John Washington, John Person. Page 202 – Vincen Harrison of CC to Robert Gill of same for 120 lbs, 100 'A' adj John Baird, Samuel Winstead, part of tract John Bridges sold to John Baird. 20 Jan 1787. Wit; Philip Vass, Gab Davey, and Buchanon.

The last two entries from Caswell County, North Carolina Deed Books are from Deed Book 'G' which records 2 more Days in the area. Isaac Day and Philip Day are listed in this 1790 transaction (CC = Caswell County).

FROM Deed Book 'G' page 143-5. William Day of Surry Co., NC to Isaac Day of CC, for 20 lbs, 173 'A' on Tar R adj. Thos Parsons, Wm Tap, W. Yarborough, Burgoon Byrd. 8 Sept 1790. Wit; James Wilson, Philip Day. Page 280 – Francis and Richard Parker of CC to John Mc Vey of same, for 150 lbs, 143 'A' on Cattail Br in Caswell & Granville Counties being a part of the old deeded land and 277 'A' of the new deeded land adj. Thomas Person, Bird's spring, Day's corner, then south to Powel Parker's corner, a total of 420 'A', 19 Jan 1791. Wit: William Yarbrough, Stephen Parker (Wm Tap = William Tapp).

FROM the 'Person County, North Carolina Deed Books' – Deed Book 'B' Page 40-1, the following records were found. Powel Parker to Richard Jones, for 27 lbs, 10 sh, 50 'A' on Tar R adj. Isaac Day, on Wagon Path from John Day to Jones; said land part of tract of Jonas Parker Decd. 10 Dec 1794. Wit: John Harris Jun., William Winsted. Page 153-4. John Day to Sion Turner of Granville Co., for 50 lbs, 200 'A' on Tar R adj. William Tapp, William Day. 4 June 1794. Wit: Robert Davie, Coleman Clayton. (Martin Cty., Halifax District). Page 167-8. Powel Parker to William Winstead, for 30 lbs, 73 'A' on Tar R adj. Richard Jones, Isaac Day at Meadow BR, John Day on Wagon Path, Bumpass old fields – it being part of tract surveyed for legatees of Jonas Parker decd. 10 Mar 1795. Wit: James Wilson, William Wilson.

These are the various types of records that are available for researching ancestors. There are many documents available, and as you can see it can be very time consuming, but very rewarding. These and many other records help in discovering where your ancestors lived, and who their neighbors were. I discovered that many of these neighbors were involved in protecting some of my ancestors.

William Howard Day

William Howard Day was born on October 19, 1825 in New York. His parents were John and Eliza Day. He worked as a printer on the Northampton Gazette before moving to Cleveland, Ohio, where he became involved in the struggle against racial discrimination. His involvement in the Abolitionist movement really began after College. After graduating from Oberlin College in 1847, he worked for the repeal of the Black Laws of Ohio. He was very concerned about:

   *   The law that required all blacks or mulattoes to carry Certificates of Freedom and to obtain two freeholders to give security for their good behavior.

   *   The law that excluded black children from attending the common schools.

   *   The law that forbade any person of color from giving testimony in a court of law against any white citizen.

William Howard Day called for a convention of Negro people in the United States in 1848. He worked diligently with Frederick Douglass for the National Negro Convention, which was held in 1849. In 1852, Day convened a large group of Colored Veterans of the War of 1812. This was the first time that these black veterans had received any public or official accolades for their service in the conflict.

While he was in college at Oberlin, Ohio Day met his wife. Her name was Lucy Stanton, and she was the first African American woman to graduate from Oberlin, College.

Inspired by the example of Frederick Douglass, Day became editor of the Cleveland True Democrat (1851-52) and the Aliened American (1853-54). In 1858, he went on a tour of Europe where he made speeches and raised funds for the Anti-Slavery cause.

Day returned to the United States after the Civil War, and worked for the Freedmen's Bureau. He became an inspector of schools in Maryland and Delaware before being ordained a minister of the African Methodist Episcopal Church in 1867. Day traveled to England, and spent a five year journey abroad. He eventually returned to America, where he founded more than 100 schools for refugees and freedmen, in Maryland, and Delaware.

William Howard Day served as general secretary of the General Conference of the African Methodist Episcopal Church (1875-1880). He resettled in Harrisburg, Pennsylvania and became elected to the school board. He died in Harrisburg on December 3rd, 1900.

Books about William Howard Day

* The African Methodist Episcopal Church, by William J. Walls.

* Men of Mark, by William J. Simmons.

* Black Americans in Cleveland, by Russell H. Davis, Assoc. Pub., (1972).

* The World of William Howard Day, by W. E. B. DuBois.

* Dictionary of American Negro Biography, by Rayford W. Logan and Michael R. Winston.

* Beating Against the Barriers, by R. J. M. Blackett (Social Science Dept. Louisiana State University, (1986).

* Black Image Makers, by Kenyette Adrine-Robinson.

* Notable Black American Men, by Jessie Carney Smith.

* Cleveland City Directory, (1861), p.31.

* From Slavery to Freedom, by John Hope Franklin, Alfred A. Knopf, Inc. (1980).

* Bishop One Hundred Years of the African Methodist Episcopal Church, by J. W. Hood, A. M. E. Zion Book Concern, (1895).

* The Necrology For Year, Oberlin College Archives, Oberlin College, (1911).

* A Documentary History of Negro People in the United States, by Herbert Aptheker, (1951).

* History of Oberlin College: From Its Foundation Through the Civil War, by Robert Samuel Fletcher, (1943).

* Black Abolitionists, by Benjamin Quarles, New York: Oxford University Press, (1969).

Books about the Day brothers - John J. Day, Jr. and Thomas J. Day

* Thomas Day, Cabinetmaker, Rodney Barfield (1975).

* An Inventory of Historic Architecture -- Caswell County, North Carolina, Ruth Little-Stokes, (1979).

* Free African Americans of North Carolina, Virginia, South Carolina: From the Colonial Period To About 1820, Volume I, Paul Heinegg, (2005).

* The Way We Lived In North Carolina, Joe A. Mobley, Editor, (2003).

* Thomas Day: African American Furniture Maker, Rodney Barfield and Patricia Marshall (2005).

* Master of Mahogany: Tom Day. Free Black Cabinetmaker (African-American Artists and Artisans), Mary E. Lyons (1994).

* Two Centuries of Black American Art, David C. Driskell, (1976).

* Dictionary of North Carolina Biography Vol. 2 (D-G), William S. Powell, Editor, (1986).

* When the Past Refused to Die: A History of Caswell County North Carolina 1777-1977, William S. Powell (1977).

* African American National Biography, Henry Louis Gates, Jr., & Evelyn Brooks Higginbotham, Eds., (2008).

* The Political and Legislative History of Liberia, Charles Henry Huberich. New York, Central Book Co., (1947).

* Black Baptists and African Missions, Stanley D. Martin, (1989).

* Slaves No More, B. I. Wiley, Ed., (1980).

* The Baptists of Virginia 1699-1926, Garnet Ryland, (1955).

* Liberia for Christ, Nan F. Weeks and Blanche Sydnor White, (1959).

* The North Carolina historical Review, Anne Miller, Editor, (2001).

* Thomas Day: Master Craftsman and Free Man of Color, Patricia Phillips Marshall & Jo Ramsay Leimenstoll, (2010).

**William A. Day, (brother of Aaron L. Day) receiving Military Awards.
He is the oldest living brother of the current generation**

APPENDIX II

Additional information about the Y- Chromosome, and the Mitochondrial DNA

Understanding the Tests Results

AFRICAN ANCESTRY

Understanding Your PatriClan™ Test Results

What is the Y chromosome?
African Ancestry studies the Y chromosome in order to determine paternal ancestry. The Y chromosome is inherited paternally. Each male has the same Y chromosome as his father, and his grandfather, and his great-grandfather, etc. Therefore, the Y chromosome provides a historical record that contains information about ancestry.

How do we read the Y chromosome?
The Y chromosome has been broken down into "markers". Markers are standard locations or addresses on the Y chromosome that allow scientists a common language when studying the DNA. Each marker is measured by its allele size. The allele size represents the number of times that a sequence pattern is repeated. For example, at marker DYS388, there is the sequence CATTG. If the allele size for marker DYS388 is 10, that means that the sequence CATTG repeats 10 times. It looks like this:

CATTG CATTG CATTG CATTG CATTG CATTG CATTG CATTG CATTG CATTG

How does the allele size affect the analysis?
Dr. Kittles reads the allele sizes for the eight markers we analyze. He compares the allele sizes to the sequences in our African Lineage Database™. When he finds a sequence that has the same allele sizes for each marker as your sequence, he finds a match. That match indicates the present-day country, and possibly ethnic group, with which you share paternal ancestry. If he doesn't find a match in the ALD, then Dr. Kittles accesses a European database and does the same analysis.

What is the YAP?
YAP stands for y alu polymorphism. Your Y chromosome either has a YAP or not. The presence of the YAP is indicated by a "+" on your printout. If you do not have a YAP, there is a "-" on your printout. Most males of African paternal descent have the YAP. Males of European descent do not. Therefore, the presence of the YAP helps Dr. Kittles with his analysis of your ancestry.

A F R I C A N   A N C E S T R Y

Understanding Your MatriClan™ Test Results

What is Mitochondrial DNA (mtDNA)?

African Ancestry studies mtDNA in order to determine maternal ancestry. MtDNA is a DNA that is only inherited maternally. We do not inherit mtDNA from our fathers. Each child (male or female) has the same mtDNA as his mother, and grandmother, and great-grandmother, etc. Therefore, mtDNA provides a historical record that contains information about maternal ancestry.

How do we read mtDNA?

We have identified a portion of your mitochondrial DNA sequence called the Hypervariable Segment I (HVSI). The sequence is broken down into base pairs, represented by the letters A, C, T, and G. We analyze 360 base pairs along the HVSI and look for mutations. This means that where the majority of humans have one base pair (or letter), there are a small number of humans that have a "different" base pair. That difference tells us something unique about your ancestry vs. everyone else. For example, let's look at ten hypothetical base pairs:

94% of the population has a "G" in position 6:	**T G T A C G G T A C**
6% of the population has an "A" in position 6:	**T G T A C A G T A C**
- This "A" is considered a mutation ⟶	

How do these mutations affect the analysis?

Dr. Kittles reads the mutations and compares them to the sequences in our African Lineage Database™. When he finds a sequence that has the same mutations as your sequence, he finds a match. That match indicates the present-day country, and possibly ethnic group, with which you share maternal ancestry. If he doesn't find a match in the ALD™, then Dr. Kittles accesses a Native American database and does the same analysis.

From the hypothetical example above, the analysis shows that this person shares ancestry with the Mende people in Sierra Leone because their sequences share an "A" in position 6. Moreover, all of the people in her family who share her maternal ancestry also share ancestry with the Mende people.

Mende person	T G T A C **A** G T A C
Woman	T G T A C **A** G T A C
Her brother	T G T A C **A** G T A C
Her mother's sister	T G T A C **A** G T A C

APPENDIX III

- The NAACP/Jefferson Leadership Academies, 'Student Project-Family History, 2006.' Attached are the names of the Students who turned in Essays, Pedigree Ancestor Sheets, and Family Group Sheets for the project.

- The 2006 & 2007 Sociology Classes at Long Beach City College in Long Beach, California. The teacher was Janét Hund, Assistant Professor, Sociology. Genealogist Aaron L. Day, presented lectures about 'Family History Preservation' at the classes, and the students prepared Pedigree Ancestor Charts, and Family Group Sheets for the project. Attached are the names of the Students who were in the classes.

Janét Hund
Assistant Professor, Sociology
jhund@lbcc.edu

- The 100 Black Men of Long Beach Charter Members. This group is working with young students, to teach them about their culture and history.

- The 'Legacy Sharing Our Past' TV Show was presented 2004 through 2006 on Long Beach Community Television - Charter Channels 65, 69, & 95. Aaron L. Day, Ernie Loewy, and Sunny Nash were co-producers of the Television show which focused on family history & genealogical research. Included are four brochures from the television shows. Day hosted the television series.

The NAACP/Jefferson Leadership Academies,

'Student Project-Family History, 2006'

Essays From The Following Students

Aguirre, Yensy	**Family Tree**
Baines, La Che	**My Beautiful Family**
Barajas, Monica	**My Family**
Davison, Brandie	**My Family**
Enamorado, Getsemani	**About Me and My Family**
Funez, Jennifer	**Family History**
Jones, Brenda C.	**My Family**
Moore, Turajha	**My Family Tree**
Mora, Emily Jewel	**The Beginning of a New Generation**
Partida, Leidy	**My Wonderful Family**
Rangel, Natalie	**My Family**
Sea, Victoria	**Family History**
Tran, Theresa	**Family Ancestors**
Verduzco, Talia	**My Family Tree**
Verduzco, Tania	**My Family Tree Essay**

<u>2006 – LBCC Sociology Class</u>

Aaron Lyday,Takreem T.

Amozoc, Juan C.

Baayoun, Sara Y.

Bandiera, Loreta R.

Buchman, Nataleigh M.

Clarke, Ronique R.

Diaz, Laura J.

Ibrahim, Sara N.

Mireles, Mariaelena.

Moore, Lakisha D.

Posada, Yolanda

Sanchez, Ivan

Valdovinos, Arley

Valencia, Javier

Wilson, Gabrielle M.

Brooks, Margie A.

Gloria, Samantha J.

Guerra, Viridiana

Matilla Lennox, Ann T.

Mehan, Melissa E.

Requeno, Maria

Vega, Gladys V.

<u>2007- LBCC Sociology Class</u>

Anaya, Guillermo

Armstrong, Michael

Breaux, Sascha K.

Brown, Taniesha S.

Burns, Jonathan E.

Butora, James

Chavis, Charles M.

Chhiv, Jonathan

Covey, Sarah A.

Cruz, Kristina M.

Cruz, Maribel L.

Davenport, William B.

Dove-Morales, Tara N.

Erickson, Emily A.

Espiritu, Jo-Anne D.

Heard, Dani R.

Henderson, Jonathon R.

Janisse, Jedon M.

Lee, Trisha M.

Lopez, Laura G.

McDaniel, Christi L.

Mendivelso, Nathaly

Merino, Stevie D.

Muinoz, Clarissa M.

Mullen, Ian N.

Negash, Erik A.

Noriega, Nathaniel E.

Oliva, Anaise M.

Padilla-Ravega, Derek A.

Richard, Cory B.

Saravia, Joel

Schafer, Brittany C.

Silva Mariscal, Lourdes

Soto, Manuela

Springs, Sundae S.

Spurlock, Brittany N.

Swanson, Coreene R.

Thomas, Rudy N.

Janét Hund - Assistant Professor, Sociology, Husband Opokuware from Kumasi, Ghana. Elder son, Jonas Kofi Oware, &Younger son, Jakobi Oware

MISSION

The mission of the *100 Black Men of Long Beach, Inc.* is to improve the quality of life within our local communities and enhance educational, health and wellness, and economic opportunities for African-Americans.

OBJECTIVE AND PURPOSE

The objective and purpose of the *100 Black Men of Long Beach, Inc.* is to foster and promote the advancement of its members in civic, economic development and charitable endeavors through mutual cooperation, joint planning, and organized execution; and to provide charitable service to the various communities of its membership so as to foster and promote the advancement of each such community.

VISION

The vision of the *100 Black Men of Long Beach, Inc.* is to be the premier community organization that serves as a catalyst to improve the quality of life for African Americans and provide the tools to become productive and community minded citizens.

GOAL

The primary goal of the *100 Black Men of Long Beach, Inc.* is to ensure the success of their primary programs as noted below with special emphsis on the mentoring program:

Mentoring Program
Health and Wellness Program
Education Program
Economic Development Program

245 West Broadway, Suite 100 · Long Beach, CA 90802

www.100blackmenlbc.org · (562) 685-0707

100 Black Men of Long Beach, Inc.

Mr. Adisa Anderson

Mr. Michael Asfall

Mr. Charles M. Brown

Mr. Ramon Brown

Mr. Raymond Brown

Mr. Eric Burroughs

Mr. Marvin R. Colburt

Mr. Aaron L. Day

Mr. Larry Dillon

Mr. Michael D. Downs

Mr. Mike Gibson

Mr. Thurman A. Green, II

Mr. Eric G. Jackson, Jr.

Mr. Jesse B. Johnson, Jr.

Mr. Ron Jones

Mr. Clarence F. Long

Mr. Ian Marcham

Mr. Todd Mason

Mr. Jason E. McCarty

Mr. John McCarty, III

Mr. Billie C. Miles, Jr.

Mr. Steve Neal

Mr. Joseph Nichols, III

Mr. Kaine Nicholas

Mr. Santos Olumese

Mr. Craig Patterson

Mr. Joel Patterson

Mr. Victor W. Parker

Mr. Howard Perry

Mr. Robin Perry

Mr. Eric Phillips

Mr. William Preston

Mr. Bill Releford, Jr.

Mr. Bill Releford, Sr.

Mr. Lance A. Robert

Mr. Ahmed C. Saafir

Mr. Raymond E. Scott, Jr.

Mr. Michael Siska, III

Mr. Darnell Smith

Mr. Eddie L. Sykes

Mr. Bert Taylor

Mr. Ashley E. Tucker

Mr. Marcus O. Tucker

Mr. Dennis Wafford

Mr. Blair R. Williams

Mr. Felton C. Williams

Mr. Edward H. J. Wilson

Mr. Tony G. Wilson

Mr. J. Nelson Woodert

Mr. Leon Woods

100 Black Men of Long Beach, Inc., Gala Inaugural, 2009
(Top-Center) Congresswoman Laura Richardson, 37th District - (Bottom - at Podium)
Dr. William Hayling - Visiting Los Angeles Members & Long Beach Members

Maycie Herrington

Celebrates the Life of Maycie Herrington

Military Historian ♦ Social Activist ♦ Family Historian

Burnett Library
560 E. Hill Street
Saturday, April 2, 2005
1:00 p.m. - 2:30 p.m.

Host, **Aaron L. Day**, with guest, **Maycie Herrington**, on the set of the *Legacy* television show

Maycie Herrington is one of two women who received a gold medal from former President George W. Bush, for serving with the Tuskeegee Airmen

Legacy

Sharing Our Past

Aaron L. Day, host and co-executive producer, is a trained genealogist, has written numerous award-winning journal articles and is the author of *Locating Free African American Ancestors.* Day consults and speaks nationally on topics that relate to family history and genealogy. **WWW.BANKSDAY.com**

Saturday Afternoon at 3:30
Long Beach Community Television
Charter Channels 65, 69 or 95

IN JULY 2004

July 10, 17, 24 & 31

Host, Aaron L. Day with guest, Evelyn Knight, on the set of *Legacy*

Legacy is multi-cultural television about family history, genealogy, community, historic sites, books and American and World History.

Ernie Loewy, executive producer of *Legacy*, is a former executive with CBS Entertainment. Loewy is the author of numerous books and consults with individuals on writing their own autobiographies.

Sunny Nash, producer of *Legacy*, is an award-winning writer, TV producer, photographer, and author of the family memoir, *Bigmama Didn't Shop At Woolworth's,* and has photographs collected by the Smithsonian Institution in Washington DC.

Consultation & Services:
- ❑ Family History & Genealogical Research Assistance
- ❑ Family Manuscript Preparation & Book Publishing
- ❑ Video Production (location shoot & editing)
- ❑ VHS & DVD Distribution
- ❑ Historic Photo Restoration, Digitizing & Disk Storage

(562) 951-9309

With her many accomplishments, Evelyn Knight is a Civil Rights Activist who marched with Dr. Martin Luther King, Jr. She is also the lead Genealogist for her family.

Aaron L. Day, host and co-executive producer, is a trained genealogist, has written numerous award-winning journal articles and is the author of *Locating Free African American Ancestors.* Day consults and speaks nationally on topics that relate to family history and genealogy. WWW.BANKSDAY.com

Saturday Afternoon at 3:30
Long Beach Community Television
Charter Channels 65, 69 or 95
September 4, 11, 18 & 25

IN SEPTEMBER!
2004

Marion Bigelow Higgins
Born June 26, 1893–the oldest person living in California

Higgins at her 111th birthday bash (left: with son, Horace and daughter-in-law, Liz; and right: with friend and assistant, Mary Gilliland)

Legacy is multi-cultural television about family history, genealogy, community, historic sites, books and American and World History.

Ernie Loewy, executive producer of *Legacy*, is a former executive with CBS Entertainment. Loewy is the author of numerous books and consults with individuals on writing their own autobiographies.

Sunny Nash, producer of *Legacy*, is an award-winning writer, TV producer, photographer, and author of the family memoir, *Bigmama Didn't Shop At Woolworth's,* and has photographs collected by the Smithsonian Institution in Washington DC.

Ripples on a Quiet Stream

The Story of the Life and Times of Marion Bigelow Higgins

Marion Higgins' book is available at
2000+ Bookstore
309 Pine Avenue, Long Beach.

Consultation & Services:

❏ Family History & Genealogical Research Assistance
❏ Family Manuscript Preparation & Book Publishing
❏ Video Production (location shoot & editing)
❏ VHS & DVD Distribution
❏ Historic Photo Restoration, Digitizing & Disk Storage

(562) 951-9309

At the airing of this Television Show, Marion Higgins, Was 111 years old - the oldest person living in California

Aaron L. Day, host and co-executive producer, is a trained genealogist, has written numerous award-winning journal articles and is the author of *Locating Free African American Ancestors*. Day consults and speaks nationally on topics that relate to family history and genealogy. **WWW.BANKSDAY.com**

Saturday Afternoon at 3:30
Long Beach Community Television
Charter Channels 65, 69 or 95

IN OCTOBER 2004

October 9, 16, 23 & 30

Bob & Marilyn Brasher have volunteered for nearly 20 years assisting others in gathering family history research at the Long Beach Public Library, where Bob received the prestigeous SPIRIT Award for excellence and dedication (lower left). Before retirement, Bob was librarian at California State University in Long Beach (lower right). Marilyn, also a former librarian, is a devoted family historian. Both have kept meticulous records and preserved many family documents, artifacts and photographs.

Distribution Assistance
Courtesy of the Long Beach
Public Library
101 Pacific Ave., Long Beach, California

Legacy **can be seen daily
on Charter Channel 63,
Courtesy of the Long Beach Public
Library**

Printing Services
Courtesy of CopyMax
4949 Lakewood Blvd., Lakewood,
California

Legacy is multi-cultural television about family history, genealogy, community, historic sites, books and American and World History.

Ernie Loewy, executive producer of *Legacy*, is a former executive with CBS Entertainment. Loewy is the author of numerous books and consults with individuals on writing their own autobiographies.

Sunny Nash, producer of *Legacy*, is an award-winning writer, TV producer, photographer, and author of the family memoir, *Bigmama Didn't Shop At Woolworth's*, and has photographs collected by the Smithsonian Institution in Washington DC.

Consultation & Services:
- ❏ Family History & Genealogical Research Assistance
- ❏ Family Manuscript Preparation & Book Publishing
- ❏ Video Production (location shoot & editing)
- ❏ VHS & DVD Distribution
- ❏ Historic Photo Restoration, Digitizing & Disk Storage

(562) 951-9309

Genealogists, Bob and Marilyn Brasher have been instrumental in helping many people locate their ancestors. They have both served on the Board of Directors for the Questing Heirs Genealogical Society of Long Beach, California for many years.

BIBLIOGRAPHY

Family Research

• *Black Genealogy*, by Charles L. Blockson and Ron Fry; Publisher: Black Classic Press; re-issue edition (April, 1997) ISBN: 0933121539

• *Finding Your People: An African-American Guide to Discovering your Roots*, by Sandra Lee Jamison; Publisher: Perigee Books; 1st edition (February, 1999) ISBN: 0399524789

• *The Right to Fight: A History of African Americans in the Military*, by Gerald Astor; Publisher: Da Capo Press (April 24, 2001) ISBN: 030681031X

• *Roots*, by Alex Haley; Publisher: Doubleday; re-issue edition (September 17, 1976) ISBN: 0385037872

• *Slaves in the Family*, by Edward Ball; Publisher: Ballantine Books (December 29, 1998) ISBN: 0345431057

• *African American Genealogy: A Bibliography for Beginners*, by Barnetta McGhee White; www.afrigeneas.com.

• *Slaves and Nonwhite Free Persons in the 1790 Federal Census of New York*, by Gilbert S. Bahn; Publisher: Clearfield Co (April, 2000) ISBN: 0806349794

• *African American Genealogical Sourcebook*, by Paula K. Byers; Publisher: Gale Group; 1st edition (April, 1995) ISBN: 0810392267

• *First Steps in Genealogy*, by Desmond Walls Allen; Publisher: Betterway Publications; 1st edition (August, 1998) ISBN: 1558704892

• *History of Black Americans, by Phillip S. Foner*

• *From Slavery to Freedom, by John Hope Franklin, New York City, New York (1947)*

• *African American History, Primary Sources, by Thomas R. Frazier (1970)*

• *The Everything Family Tree Book, by Holbrook (1998)*

• *The Sweeter The Juice, by Shirlee Taylor Haizlip, New York City, New York (1944)*

• *Northumberland County, Virginia Apprenticeships, 1650-1750 by Preston W. Haynie*

• *Northumberland County, Virginia Apprenticeships, 1750-1852 by Preston W. Haynie*

• *Records of indentured Servants and of Certificates for Land, Northumberland County, Virginia 1650-1795, by Preston W. Haynie*

• *Netting Your Ancestors, by Cyndi Howells, Baltimore, Maryland (1997)*

• *A Student's Guide To African American Genealogy, by Anne E. Johnson, and Adam Merton (1996)*

• *Caswell County Apprentice Bonds, by Katherine Kerr Kendall, Raleigh, North Carolina (1984)*

• *Person County Marriage Bonds, by Katherine Kerr Kendall, Person County, North Carolina (1983)*

• *The Sourcebook Guidebook of American Genealogy, by Loretto D. Szucs, and Sandra Hargreaves Luebking*

• *Case Studies In Afro-American Genealogy, by David T. Thackery, and Dee Woodtor, Chicago, Illinois (1989)*

• *Do People Grow On Family Trees? By Ira Wolfman, New York City, New York (1997)*

About - Free African Americans

1. *Register of Free Blacks: Rockingham County, Virginia, 1807-1859*, by Dorothy A. Boyd-Bush; Publisher: Heritage Books; (October, 1992), ISBN: 1556136587

2. *Free Blacks in America, 1800-1860*, edited by John H. Bracey, Jr., August Meier, and Elliott Rudwick; Publisher: Wadsworth Pub. Co.; - (1971), ISBN: 0534000223

3. *Master of Mahogany – Tom Day, Free Black Cabinetmaker*, by Mary Lyons; Publisher: New York Scribner's.

4. *Pittsylvania County, Virginia Register of Free Negroes*, by Alva H. Griffith; Publisher: Heritage Books Inc; (January, 2001), ISBN: 0788417800

5. *The Free Negro Family : A Study of Family Origins Before the Civil War*, by Edward Franklin Frazier; Publisher: Beaufort Books; Reprint edition; (May, 1979)
ISBN: 0405018150

6. *Creole: The History and Legacy of Louisiana's Free People of Color*, edited by Sybul Kein; Publisher: Louisiana State University Press; (2000), ISBN 0-8071-2601-2

7. *Free Blacks in Hartford, Somerset, and Talbot Counties, Maryland, 1832*, compiled by Mary K. Meyer; Publisher: Pipe Creek Publications; (1991).

8. *Slaves and Nonwhite Free Persons in the 1790 Federal Census of New York*, by Gilbert S. Bahn; Publisher: Clearfield Co; (April, 2000), ISBN: 0806349794

9. *Strangers In Their Midst: The Free Black Population of Amherst County, Virginia*, by Sherrie S. McLeroy, William R. McLeroy; Publisher: Heritage Books; (June, 1993)
ISBN: 1556137869

10. *Against The Odds: Free Blacks In The Slave Societies Of The Americas*, edited by Jane G. Landers; Publisher: Frank Cass Publishers; (May, 1996), ISBN: 0714642541

11. *Free Black Heads of Households in the New York State Federal Census, 1790-1830*, by Alice Eichholz (Editor); Publisher: Gale Group; (March, 1981), ISBN: 0810314681

12. *Register of Free Negroes and also of Dower Slaves, Brunswick County, Virginia, 1803-1850*, transcribed by Frances Holloway Wynne; Publisher: F.H. Wynne; (1983)
ASIN: B0006EJ1BY

13. *Free Blacks of Anne Arundel County, Maryland, 1850*, by Ralph Clayton; Publisher: Heritage Books; (June, 1987), ISBN: 1556130694

14. *Neither Slave nor Free: The Freedmen Of African Descent In The Slave Societies Of The New World*, edited by David W. Cohen and Jack P. Greene; Publisher: The Johns Hopkins University Press; (September 1, 1974), ISBN: 0801816475

15. *The Register of Free Negroes, Northampton County, Virginia, 1853 to 1861*, transcribed, referenced, and indexed by Frances Bibbins Latimer; Publisher: Heritage Books; (1992)
ISBN: 1556136226

16. *The Free Negro in Ante-Bellum Louisiana*, by Herbert E. Sterkx; Publisher: Associated Univ. Press; (June, 1972), ISBN: 0838678378

17. *The Free Negro In Texas, 1800-1860: A study in Cultural Compromise*, by George Ruble Woolfolk; Publisher: Published for the Journal of Mexican American History by University Microfilms International; (1976), ISBN: 0835701921

18. *The Free Negro Owners of Slaves in the United States in 1830*, by Carter G. Woodson; Publisher: Greenwood Press Reprint; (April 30, 1969), ISBN: 083710761X

19. *From Slavery To Freedom: A History Of Negro Americans*, by John Hope Franklin; Publisher: Alfred a Knopf; 5th edition; (June, 1978), ISBN: 0394322568

20. *Free Women of Petersburg: Status and Culture in a Southern Town, 1784-1860*, by Suzanne Lebsock; Publisher: W. W. Norton & Company; New Ed edition; (October, 1985)
ISBN: 0393952649

21. *Free Blacks and Mulattos in South Carolina 1850 Census*, by Margaret Peckham Motes; Publisher: Clearfield Co (October, 2000), ISBN: 0806350261

DNA Books/Websites

• DNA & Genealogy, by Colleen Fitzpatrick and Andrew Yeiser,
www.forensicgenealogy.info

• Forensic Genealogy, by Colleen Fitzpatrick, www.forensicgenealogy.info

• Trace Your Roots With DNA, by Megan Smolenyak- Smolenyak, and Ann Turner,
www.rodalestore.com

• DNA & Family History, by Chris Pomery, www.DNAandfamilyhistory.com

• Scientists Who Made History: Rosalind Franklin, by Cath Senker, Raintree Steck-
Vaughn Publishers, Austin, Texas

• Oswald Avery: And The Story Of DNA, by Vesta-Nadine Severs and Jim Whiting,
Mitchell Lane Publishers

• The Human Genome Project: What Does Decoding DNA Mean For Us? By Kevin
Alexander Boon, Enslow Publishers, Inc.

• How To DNA Test Our Family Relationships, by Terrence Carmichael, and Alexander
Kukin, Ace N Press, Mountain View, California, ISBN 0-9664027-1-5

• How Did We Find Out About DNA? By Isaac Asimov, Walker and Company, New York,
NY

• Finding Oprah's Roots, by Henry Louis Gates, Jr., Crown Publishers, New York

• Access To The Genome: The Challenge To Equality, by Maxwell J. Mehlman and
Jeffrey R. Botkin, Georgetown University Press, Washington, DC

• Rosalind Franklin, www.curie.che.virginia.edu

• The Tech: Rosalind Franklin, www.thetech.org

• Microbe Hunters, by Paul De Kruif, Pleasantville, New York, Readers Digest
Association, features Oswald Avery (1962)

• The Professor, the Institute, and DNA, Rene Dubois, New York, The Rockefeller
University Press (1956)

• Actual Innocence, by Jin Dwyer, Barry Scheck, and Peter Neufield, New York:
Doubleday (2000)

• The Hoagland Laboratory, by Arnold H. Eggerth, Brooklyn, New York (1960)

- DNA: The Thread Of Life, by Frank H. Wilcox, Lerner Publications Company (1988)

- Human Genome Project Information, www.ornl.gov

- National Institute of Health, National Human Genome Research Institute, www.nhgri.nih.gov

- Wikipedia Free Library -   www.wikipedia.org

(Search)   (DNA)

Africa - People/Countries/Cultures

- Discovering Africa's Land, People, and Wildlife, by Aileen Weintraub, Enslow Publishing, Inc. (2004) www.myreportslinks.combooks

- Discovering Cultures: Nigeria, by Patricia J. Murphy, Benchmark Books, Marshall Cavendish, New York (2005)

- Nigeria The People, by Anne Rosenberg, a Bobbie Kalman book, The Lands, Peoples, and Cultures Series, Crabtree Publishing Company (2001)

- Griots and Griottes, by Thomas A. Hale, Indiana University Press, Bloomington and Indianapolis (1998)

- Oral Literature In Africa, by Ruth Finnegan, the Oxford Library of African Literature, Oxford at the Clarendon Press (1970)

- Cultures of the World: Nigeria, by Patricia Levy, Benchmark Books, Marshall Cavendish, New York

- Countries of the World: Liberia, by Paul Rozario, Gareth Stevens Publishing, A World Almanac Education Group Company

- The Kingdoms of Africa (The Making of the Past Series), by Peter Garlake, Peter Bedrick Books, New York (1978)

- The Tree Where Man Was Born: The African Experience, by Peter Matthiessen and Eliot Porter, E. P. Dutton & Co., Inc., New York (1972)

- Black Kingdom Black peoples: The West African Heritage, by Anthony Atmore and Gillian Stacey, Golden and Golden Press (1979)

- Country Fact Files: East Africa, by Rob Bowden and Tony Binns, Raintree Steck-Vaughn Publishers, Austin, Texas (1998)

- Country Fact Files: West Africa, by Tony Binns and Rob Bowden, Raintree Steck-Vaughn Publishers, Austin, Texas (1998)

- Nations of the World: South Africa, by Jen Green, Raintree Steck-Vaughn Publishers, Austin , Texas (2001)

- Continents: Africa, by Colm Regan and Pedar Cremin, Raintree Steck-Vaughn Publishers, Austin, Texas (1997

- Cultures of the World: Ghana, by Patricia Levy, Marshall Cavendish Corporation, New York, London, & Sydney (1999)

- Countries of the World: Kenya, by Victoria Derr, Gareth Stevens publishing, Milwaukee (1999)

- African Religion: World Religions, by Aloysius M. Lugira, Facts On File, Inc., New York, NY 2004, 1999

INDEX

There were no entries for **X and Z**

Made in the USA
Monee, IL
07 July 2026

56564663R00142